Regenerative Medicine

A comprehensive review of the challenges that exist in patient accessibility to regenerative medicines (RMs), presenting clinical trials, marketing authorization, HTA, pricing, reimbursement, affordability, payment and partnership agreements of RMs, and commercialization. Specifically, we investigate how COVID-19 has impacted the RM industry by elaborating on the disruptions it caused but also the new opportunities it brought. The ultimate goal of this work is to make strategic recommendations for manufacturers and decision-makers on effective strategies to address these obstacles and facilitate patient access to promising regenerative medicines.

FEATURES

Regenerative medicine (RM) is an emerging interdisciplinary field aiming to replace or regenerate human cells, tissues, or organs in order to restore normal function.

1. RM holds the promise of revolutionizing treatment in the 21st century.
2. RMs bring new hope for some previously untreatable diseases, as well as holding promise for the treatment of common chronic diseases.
3. Rapid advancements in biotechnology and improved understanding of disease pathophysiology have attracted tremendous interests in the development of RMs.
4. Discusses the high cost of RMs, which may challenge the sustainability of healthcare insurers (public and private).

Pharmaceuticals, Health Economics and Market Access

Editor:
Mondher Toumi

Gene and Cell Therapies: Market Access and Funding
Eve Hanna and Mondher Toumi

Managed Entry Agreements and Funding for Expensive Therapies
Mondher Toumi, Szymon Jaroslawski

Regenerative Medicine: Unlocking Patient Access and Commercial Potential
Tingting Qiu and Mondher Toumi

For more information about this series, please visit: www.crcpress.com/Pharma
ceuticals-Health-Economics-and-Market-Access/book-series/CRCPHEMA

Regenerative Medicine
Unlocking Patient Access and Commercial Potential

Tingting Qiu
Beijing Institute of Clinical Pharmacy,
Beijing Friendship Hospital of Capital Medical University,
Beijing, China
Mondher Toumi
Aix Marseille University, France

CRC Press
Taylor & Francis Group
Boca Raton London New York

CRC Press is an imprint of the
Taylor & Francis Group, an **informa** business

First edition published 2023
by CRC Press
6000 Broken Sound Parkway NW, Suite 300, Boca Raton, FL 33487–2742

and by CRC Press
4 Park Square, Milton Park, Abingdon, Oxon, OX14 4RN

© 2023 Tingting Qiu and Mondher Toumi.

CRC Press is an imprint of Taylor & Francis Group, LLC

ISBN: 978-1-032-43296-0 (HB)
ISBN: 978-1-032-43198-7 (PB)
ISBN: 978-1-003-36667-6 (EB)

DOI: 10.1201/9781003366676

Typeset in Times
by Apex CoVantage, LLC

Contents

Preface

Regenerative medicine (RM) is an emerging interdisciplinary field aiming to replace or regenerate human cells, tissues, or organs in order to restore normal function (1). Not only may these therapies relieve disease symptoms, but they have also shown great potential to prevent the onset of disease and delay or halt disease progression. RMs bring new hope for some previously untreatable diseases, as well as holding promise in the treatment of common chronic diseases, such as stroke, Parkinson disease, and renal failure (2).

Rapid advancements in biotechnology and improved understanding of disease pathophysiology have attracted tremendous interests in the development of RMs. Important acquisitions of RMs by pharmaceutical companies have been seen in the past 5 years (3). According to the annual report of the Alliance of Regenerative Medicine (ARM), more than 1,000 clinical trials investigating RMs for the treatment of various diseases are underway as of the third quarter of 2019 (4). As of February 2020, it was estimated that RMs have accounted for 12% of the industry's clinical pipeline based on the manufacturers' disclosures of their assets, implying the huge commercial potential of RMs in the future (3). Globally, more than 50 RMs have been approved by February 2022, with the United States and Europe at the forefront, followed by Asian countries, including Japan, China, South Korea, and India (5).

Despite rapid development and the great clinical potential of RMs, compared to traditional medicines and biologicals, RMs face significant difficulties in transiting scientific discovery to successful commercialization (1). To begin with, the manufacturing and quality control of advanced therapy medicinal products (ATMPs) are complex processes, which must be carefully designed to guarantee quality, stability, consistency, and safety. For example, the isolation of viable cells from patients to obtain the cellular starting materials is time-consuming, and the current process for vector production is still inefficient despite significant progress (2). In the clinical development stage, reasonably sized, randomized control trials (RCTs) seem infeasible for ATMPs due to small patient numbers, the absence of effective treatments to serve as active comparators, and the ethical controversy generated from assigning fragile patients to placebo when no effective treatment is available (3). The lottery of assigning patients to a placebo for a severe disabling or life-threatening progressive irreversible disease while a new therapy with potential to halt disease progression or cure the disease exists raised a lot of controversy. Clinicians and patients alike expressed reluctance to engage in such trials.

RMs may represent major therapeutic promises, making them likely to be eligible for expedited access pathways, such as the Regenerative Medicine Advanced Therapy (RMAT) and Breakthrough Therapy designation in the United States, approval under exceptional circumstances, conditional approval, and PRIority MEdicines (PRIME) in the European Union (EU) (5). However, in contrast to regulators, health technology assessment (HTA) bodies hold more conservative attitudes towards RMs. They concluded that limited clinical evidence made reliable assessment of the effectiveness, safety, and economic outcomes of RMs highly uncertain or even impossible (6). The high cost of RMs may challenge the sustainability of healthcare insurers (public and private). Innovative payment mechanisms, such as outcome-based payment and installment payment, have been proposed to mitigate the financial risk of paying for products that have limited evidence of effectiveness and safety at the time of launch. However, the feasibility of implementing such payment models into existing health insurance systems has proven challenging (7). All these challenges related to the market access of RMs became even more paramount during the COVID-19 pandemic crisis when patients were concerned with the risk of exposure to COVID-19, healthcare system faced a shortage of medical resources, and payers had to allocate their constricted budgets to an emergency basis.

Therefore, in this book, we aim to first provide a comprehensive review of all the challenges existing in patients' accessibility of RMs, starting with clinical trials, marketing authorization, HTAs, pricing, reimbursement, affordability, payment, and partnership agreements of RMs. Specially, we investigate how COVID-19 has impacted the RM industry by elaborating on the disruptions it caused but also the new opportunities it brought. The ultimate goal of this work is to make strategic recommendations for manufacturers and decision-makers on effective strategies to addressing all these obstacles to enable facilitated patient access to promising RMs that are widely expected by patients and healthcare professionals.

Author Biographies

Tingting Qiu holds a PhD degree in public health and a master's degree in pharmacology. Her area of expertise includes regulatory affairs, health policy, health technology assessment, economic analysis, and price and reimbursement of innovative therapies, such as orphan drugs and regenerative medicines. She has authored over 30 publications and conference presentations focused on the market access of cell and gene therapies at the European and global level. She worked as a price and market access analyst at Creativ-Ceutical (Paris, France) for 4 years from 2018, during which she was deeply involved in addressing the challenges in meeting the expectations of relevant stakeholders to facilitate patient access to lifesaving treatments. Since 2021, she has been a research associate to EA 3279 CERESS Department of Public Health at Aix-Marseille University. Currently, she is a researcher in the Beijing Institute of Clinical Pharmacy, Beijing Friendship Hospital, Capital Medical University (Beijing, China). Apart from endeavoring to advance market access of cell and gene therapies in China and globally, she is also engaged in projects focusing on the impact of implementing diagnosis-related groups reform on the healthcare system, the methodology of health economic analysis in China, and evidence-based assessment of Chinese and western medicines.

Mondher Toumi is an MD by training and has an MSc in biostatistics and in biological sciences (option pharmacology) and a PhD in economic sciences. Toumi is a professor of public health at Aix-Marseille University. After working for 12 years as a research manager in the Department of Pharmacology at the University of Marseille, he joined the Public Health Department in 1993. In 1995 he embraced a career in the pharmaceutical industry for 13 years. Toumi was appointed global vice president at Lundbeck A/S in charge of health economics, outcome research, pricing, market access, epidemiology, risk management, governmental affairs, and competitive intelligence. In 2008, he founded Creativ-Ceutical, an international consulting firm dedicated to supporting health industries and authorities in strategic decision-making. In February 2009 he was appointed professor at Lyon I University in the Department of Decision Sciences and Health Policies. The same year, he was appointed director of the Chair of Public Health and Market Access. He launched the first European University Diploma of Market Access (EMAUD), an international course currently followed by almost 350 students. Additionally, he recently created the Market Access Society to promote education, research, and scientific activities at the interface of market access, HTA, public health, and health economic assessment. He is editor-in-chief of *Journal of Market Access and Health Policy* (JMAHP), which was just granted PubMed indexation. Toumi is also a visiting professor at Beijing University (Third Hospital). He is a recognized expert in health economics and an authority on market access and risk management. He has more than 250 scientific publications and communications and has contributed to several books.

Abbreviations

AAV	Adeno-associated virus
ARM	Alliance for Regenerative Medicine
ATMP	Advanced therapy medicinal product
BTD	Breakthrough designation
CAR	Chimeric antigen receptor
CAT	Committee for Advanced Therapies
CBER	Center for Biologics Evaluation and Research
CDF	Cancer Drug Fund
CEA	Cost-effectiveness analysis
CHMP	Committee for Medicinal Products for Human Use
COVID	Coronavirus disease
CRISPR	Clustered regularly interspaced short palindromic repeats
DLBCL	Diffuse large B-cell lymphoma
DNA	Deoxyribonucleic acid
DRG	Diagnosis-related group
EBMT	European Society for Blood and Marrow Transplantation
EMA	European Medicines Agency
EU	European Union
EUR	Euro (€)
FDA	Food and Drug Administration
FTD	Fast-track designation
GCP	Good clinical practice
HAS	Haute Autorité de Santé
ICER	Incremental cost-effectiveness ratio
ICH	International Council for Harmonization
LPLD	Lipoprotein lipase deficiency
MFDS	Ministry of Food and Drug Safety
MHLW	Ministry of Health, Labour and Welfare
NCT	Number of clinical trials
OD	Orphan drug
ODD	Orphan drug designation
OS	Overall survival
PMBCL	Primary mediastinal large B-cell lymphoma
QOL	Quality of life
RM	Regenerative medicine
RMAT	Regenerative medicine advanced therapy
RWE	Real-world evidence
USA	United States of America
WHO	World Health Organization

1 Regulation of RMs in Globally

Despite the rapid growth in the research and clinical trials of RM, only a small number have been approved at a global level. Regulatory hurdles constitute one of the biggest challenges for successful clinical translation and commercialization of RMs (8). Bravery et al (9) observe that RMs, especially non-orphan RMs, are associated with a lower approval rate and take longer to approve compared with other products. MA requirements, such as compliance with good manufacturing procedures (GMPs) and conducting pivotal studies with robust study design and large enough patient populations for efficacy and safety investigation, appear to be too burdensome to fulfill and have failed to consider the unique characteristic of RMs (10). Contrary to the slow progress in the MA approval of RMs, the use of unauthorized RMs prepared in private clinics has continuously increased in some countries, raising significant concerns around their potential safety risk (11).

Regulators have therefore realized the significance of standardizing the approval processes of RMs to prevent patients from potential risks from unapproved and/or untested treatments. Since 2007, RM legislation outlining the regulatory requirements have begun to be established in several countries. In this chapter, we intend to provide a global picture on the regulation of RMs in terms of MA pathways, characteristics of RMs, and accelerated regulatory pathways in each country. By providing information on the regulatory policies implemented in the investigated countries and the EU, further development and strengthening of the competencies in RMs may be facilitated.

1.1 REGULATION OF RMS IN EACH COUNTRY

1.1.1 EUROPEAN UNION (EU)

1.1.1.1 Regulation on Advanced Therapy Medicinal Products

Regulation (EC) no. 1394/2007 was released on 13 November 2007 as a specific legislative act to regulate the whole lifecycle of advanced therapy medicinal products (ATMPs) from the drug development phase to the ultimate post-market surveillance. According to this regulation, ATMPs should be subject to the same regulatory principles as other types of biotechnology medicinal products. However, technical requirements, in particular the nature and volume of quality, preclinical, and clinical data necessary to demonstrate the quality, safety, and efficacy of the product, may be highly specific and flexible.

Afterwards, a set of scientific guidelines specific to ATMPs were issued, such as the guideline on the quality, non-clinical, and clinical aspects of gene therapy

DOI: 10.1201/9781003366676-1

medicinal products (CHMP/GTWP/671639/2008), guideline on human cell–based medicinal products (EMEA/CHMP/410869/2006), guideline on safety and efficacy follow-up and risk management of advanced therapy medicinal products (EMEA/149995/2008), and guidelines on good manufacturing practice specific to advanced therapy medicinal products (4). Additionally, the application of a risk-based approach in the Marketing Authorization Application (MAA) dossier was encouraged to help manufacturers determine what extent of quality, non-clinical, and clinical data should be included in the MAA in accordance with the scientific guidelines published by the EMA.

Exceptional products exempted from Regulation (EC) no. 1394/2007 were defined as products under the "hospital exemption" rule, and eligible ATMPs must satisfy the following criteria: prepared on a non-routine basis according to specific quality standards and utilized within the same member state in a hospital under the exclusive professional responsibility of a medical practitioner to comply with an individual medical prescription for a custom-made product for an individual patient.

1.1.1.2 Market Authorization of ATMPs

ATMPs are subject to a centralized market authorization procedure, which was expected to ensure a high-quality scientific evaluation. After receiving the request for market authorization, a Committee for Advanced Therapies (CAT) will evaluate the dossiers submitted by the manufacturers; deliver scientific recommendations; and present draft opinions regarding the quality, safety, and efficacy to the CHMP. The CHMP recommendation will then be sent to the European Commission for final approval under the current standard procedure for human medicinal products.

Medicines for the treatments of diseases with major public health interest or high unmet medical needs could be entitled to conditional market authorization, but MA holders of conditional market authorization must fulfill a specific obligation to further collect post-market evidence on the efficacy and safety of medicines before a standard market authorization is obtained. An adaptive pathway approach based on the principle of iterative development, real-world evidence collection, and early involvement of multiple stakeholders was implemented with the objective to accelerate patient access to innovative medicines with urgent needs. The request for accelerated assessment for ATMP could be sent to CAT; once granted, the assessment timeframe for the MA application will reduce to 150 days rather than 210 days under the standard procedure. The PRIority MEdicines (PRIME) scheme was launched in March 2016 to enhance the market authorization of medicines with major therapeutic advantages over existing treatments or substantial benefits to patients without treatment options. Although such strategies were applicable for all innovative medicines, ATMPs indicated for previously untreatable diseases with limited treatment alternatives were highly likely to obtain market authorization through the previously mentioned expedited approach.

1.1.1.3 ATMPs Approved in the EU

As of March 2022, a total of 20 ATMPs were approved in the EU (Table 1.1), among which, most of them were gene therapies for oncology diseases. Fifteen out

TABLE 1.1

Regenerative Medicines Approved in the European Union

Name	ATMP Type	Indication	Therapeutic Area	MA Date	Designation	MA Pathway
ChondroCelect (characterized by viable autologous cartilage cells expanded ex vivo expressing specific marker proteins)	Tissue engineered product	Used in adults to repair damage to the cartilage in the knee	Musculoskeletal diseases	05/10/2009	N/A	None
Glybera (alipogene tiparvovec)	Gene therapy	Adults with lipoprotein lipase deficiency	Lipid-modifying agents	25/10/2012	Orphan designation	Approved under EC
Provenge (autologous peripheral-blood mononuclear cells)	Somatic cell therapy product	Prostatic neoplasms	Oncologic disease	06/09/2013	N/A	None
MACI (matrix-applied characterized autologous cultured chondrocytes)	Tissue engineered product	Repair full-thickness defects with a surface area of between 3 and 20 cm^2 in adults who are experiencing symptoms (such as pain and problems moving the knee).	Musculoskeletal diseases	27/06/2013	N/A	None
Imlygic (talimogene laherparepvec)	Gene therapy	Melanoma	Oncologic disease	16/12/2015	N/A	None
Holoclar (ex vivo expanded autologous human corneal epithelial cells)	Tissue engineered product	Corneal lesions, with associated corneal (limbal) stem cell deficiency (LSCD), due to ocular burns.	Optical disease	17/02/2015	Orphan designation	Conditional approval
Zalmoxis (allogeneic T cells genetically modified)	Somatic cell therapy product	Patients receiving haploidentical HSCT	Oncologic adjuvant therapy	18/08/2016	Orphan designation	Conditional approval;
Strimvelis (autologous CD34+ enriched cell fraction)	Gene therapy	Severe combined immunodeficiency due to adenosine deaminase deficiency (ADA-SCID)	Enzyme deficiency	26/05/2016	Orphan designation	None

(*Continued*)

TABLE 1.1 (Continued)
Regenerative Medicines Approved in the European Union

Name	ATMP Type	Indication	Therapeutic Area	MA Date	Designation	MA Pathway
Spherox (spheroids of human autologous matrix-associated chondrocytes)	Tissue engineered product	Articular cartilage defects	Musculoskeletal diseases	10/07/2017	N/A	None
Luxturna (voretigene neparvovec)	Gene therapy	Vision loss	Ophthalmologic disease	22/11/2018	Orphan designation	None
Yescarta (axicabtagene ciloleucel)	Gene therapy	• DLBCL • PMBCL	Oncologic disease	23/08/2018	Orphan designation; PRIME designation	None
Kymriah (tisagenlecleucel)	Gene therapy	• DLBCL • ALL	Oncologic disease	22/08/2018	Orphan designation; PRIME designation	None
Alofisel (darvadstrocel)	Somatic cell therapy product	Perianal fistulas	GI disease	23/03/2018	Orphan designation;	None
Zynteglo (autologous CD34+ cells encoding the βA-T87Q-globin gene)	Gene therapy	Beta -thalassemia in patients 12 years and older who require regular blood transfusions.	Hematologic disease	29/05/2019	Orphan designation; PRIME designation	None
Zolgensma (onasemnogene abeparvovec-xioi)	Gene therapy	Spinal muscular atrophy (SMA)	Neurologic disease	18/05/2020	Orphan designation;	Conditional approval

Tecartus (brexucabtagene autoleucel)	Gene therapy	Mantel cell lymphoma; acute lymphocytic leukemia	Oncologic disease	14/12/2020	Orphan designation; Conditional approval
Libmeldy (atidarsagene autotemcel)	Gene therapy	Metachromatic leukodystrophy	Metabolism disorder	17/12/2020	Orphan designation; None
Abecma (idecabtagene vicleucel)	Gene therapy	Multiple myeloma	Oncologic disease	18/08/2021	Orphan designation; Conditional approval
Skysona (elivaldogene autotemcel)	Gene therapy	Adrenoleukodystrophy	CNS disease	16/07/2021	Orphan designation; None
Breyanzi (lisocabtagene maraleucel)	Gene therapy	Diffuse large B-cell lymphoma; follicular lymphoma	Oncologic disease	27/01/2022	Orphan designation; None

of 20 products were orphan drugs, and 6 out of 20 products received conditional approval or approval under exceptional circumstances. Seven products were withdrawn from the EU market, including ChondroCelect, Glybera, Provenge, MACI, Zalmoxis, Zynteglo, and Skysona.

1.1.2 UNITED STATES

1.1.2.1 Regulation on Cell and Gene Therapies

Regenerative medicine was regulated by the Center for Biologics Evaluation and Research (CBER) using the existing regulations for drugs and biological products, that is, the Public Health Service Act (PHS Act) and the federal Food, Drug and Cosmetic Act (FD&C Act). The Division of Cell and Gene Therapies in the Office of Tissue and Advanced Therapies (OTAT) conducted the mission of evaluating regulatory files associated with cell and gene therapies.

Regenerative medicine was considered to involve "human cells, tissues, or cellular or tissue-based products" (HCT/Ps) in Title 21 of the Code of Federal Regulations (CFR) Part 1271.3(d) (21 CFR 1271). A tiered, risk-based approach for HCT/Ps regulation was proposed and implemented by the FDA in 2005 based on the potential risk related to the transmission of communicable diseases, necessary processing controls to prevent contamination, and preservation of the integrity and original function, as well as the efficacy and safety of products. Regenerative medicines could be generally divided into two main categories, known as "Section 361" and "Section 351" HCT/Ps under the PHS Act. The considerable distinction between "Section 351" and "Section 361" HCT/Ps existed in terms of definition and regulatory requirements. HCT/Ps that fall into Section 361 will be subject solely to regulation under Section 361 of the PHS Act and 21 CFR Part 1271, which include establishment registration and product listing, assessment of donor eligibility, and compliance with Good Tissue Practice (cGTP), but pre-marketing approval is not required. Section 361 HCT/Ps only apply to products that are minimally manipulated (Box 1.1), intended for homologous use, non-combined products, and not associated with systematic effects and not dependent upon the metabolic activity of living cells for their primary functions. If the HCT/Ps do not meet the requirements of Section 361 HCT/Ps, they will be regulated as drugs or biological products under Section 351 of the PHS Act and the FD&C Act, such as compliance with cGMP and obtaining pre-market approval based on clinical trial data. As a part of the FDA's comprehensive regenerative medicine policy framework, a final guideline document called "Regulatory Consideration for Human Cells, Tissues, and Cellular and Tissue-Based Products: Minimal Manipulation and Homologous Use" was published in November 2017 to clarify the principles used to determine the classification of HCT/Ps and whether the pre-market review requirements need to be fulfilled. A more detailed list of the guidelines on regenerative medicine published since 2017 is provided in Table 1.2. The series of newly published guidelines covered not only the general recommendations on the manufacturing, testing, and clinical design for regenerative medicines

> **BOX 1.1   THE DEFINITION OF MINIMAL MANIPULATION FROM FDA GUIDELINES**
>
> Section 1271.3(f) provides two definitions of minimal manipulation, one that applies to structural tissue and one that applies to cells or non-structural tissues.
>
> For structural tissue, minimal manipulation means that the processing of the HCT/P does not alter the original relevant characteristics of the tissue relating to its utility for reconstruction, repair, or replacement [21 CFR 1271.3(f)(1)]. For cells or non-structural tissues, minimal manipulation means that the processing of the HCT/P does not alter the relevant biological characteristics of cells or tissues [21 CFR 1271.3(f)(2)].
>
> *Source*: FDA guideline: Regulatory Considerations for Human Cells, Tissues, and Cellular and Tissue-Based Products: Minimal Manipulation and Homologous Use

TABLE 1.2
Guidelines on FDA Regulations on Regenerative Medicines

Title	Objective of the Guideline	Date of Issue
Considerations for the Development of Chimeric Antigen Receptor (CAR) T Cell Products	Provide CAR T-cell–specific recommendations regarding chemistry, manufacturing, and control (CMC); pharmacology; toxicology; and clinical study design	March 2022
Human Gene Therapy Products Incorporating Human Genome Editing	Provides recommendations regarding information that should be provided in an Investigational New Drug (IND) application to assess the safety and quality of the investigational genome editing (GE) product. This includes information on product design, product manufacturing, product testing, pre-clinical safety assessment, and clinical trial design.	March 2022
Interpreting Sameness of Gene Therapy Products Under the Orphan Drug Regulations	Focuses specifically on factors that the FDA generally intends to consider when determining sameness for gene therapy products and does not address sameness determinations for other types of products.	September 2021

(Continued)

TABLE 1.2 *(Continued)*
Guidelines on FDA Regulations on Regenerative Medicines

Title	Objective of the Guideline	Date of Issue
Studying Multiple Versions of a Cellular or Gene Therapy Product in an Early-Phase Clinical Trial	Provides recommendations for studies that evaluate multiple versions of a cellular or gene therapy product, including how to organize and structure the INDs, submit new information, and report adverse events.	September 2021
Manufacturing Considerations for Licensed and Investigational Cellular and Gene Therapy Products During COVID-19 Public Health Emergency	Provides manufacturers of licensed and investigational cellular therapy and gene therapy (CGT) products with risk-based recommendations to minimize potential transmission of the novel coronavirus, severe acute respiratory syndrome coronavirus 2 (SARS-CoV-2).	January 2021
Human Gene Therapy for Neurodegenerative Diseases	Focuses on considerations for product development, pre-clinical testing, and clinical trial design. The guidance also discusses marketing approval pathways for investigational gene therapy products.	January 2021
Chemistry, Manufacturing, and Control (CMC) Information for Human Gene Therapy Investigational New Drug Applications (INDs)	Informs sponsors how to provide sufficient CMC information required to assure product safety, identity, quality, purity, and strength (including potency) of the investigational product.	January 2020
Long Term Follow-up After Administration of Human Gene Therapy Products	Provides recommendations regarding the design of long-term follow-up studies (LTFU observations) for the collection of data on delayed adverse events following administration of a gene therapy (GT) product.	January 2020
Testing of Retroviral Vector-Based Human Gene Therapy Products for Replication Competent Retrovirus During Product Manufacture and Patient Follow-up	Provides recommendations regarding the testing for replication competent retrovirus (RCR) during the manufacture of retroviral vector-based gene therapy products and during follow-up monitoring of patients who have received retroviral vector-based gene therapy products.	January 2020

TABLE 1.2 *(Continued)*
Guidelines on FDA Regulations on Regenerative Medicines

Title	Objective of the Guideline	Date of Issue
Human Gene Therapy for Hemophilia	Provides recommendations to sponsors developing human GT products for the treatment of hemophilia, including clinical trial design and related development of coagulation factor VIII (hemophilia A) and IX (hemophilia B) activity assays, including how to address discrepancies in factor VIII and factor IX activity assays.	January 2020
Human Gene Therapy for Rare Diseases	Provides recommendations to sponsors developing human GT products intended to treat a rare disease in adult and/or pediatric patients regarding the manufacturing, pre-clinical, and clinical trial design issues for all phases of the clinical development program.	January 2020
Human Gene Therapy for Retinal Disorders	Provides recommendations to sponsors developing human GT products for retinal disorders affecting adult and pediatric patients. These disorders vary in etiology, prevalence, diagnosis, and management and include genetic as well as age-related diseases.	January 2020
Evaluation of Devices Used with Regenerative Medicine Advanced Therapies	Provides manufacturers, applicants, and sponsors engaged in the development of regenerative medicine therapies the FDA's current thinking regarding the evaluation of devices used in the recovery, isolation, or delivery of regenerative medicine advanced therapies.	February 2019
Expedited Programs for Regenerative Medicine Therapies for Serious Conditions	Describes the expedited programs available to sponsors of regenerative medicine therapies for serious conditions, including those products designated as RMAT.	February 2019

but also the recommendations on the specific disease areas, including rare diseases, retinal diseases, neurological diseases, and hemophilia.

1.1.2.2 Expedited Pathway to Accelerate the Market Authorization

The final guideline "Expedited Programs for Regenerative Medicines Therapy for Serious Conditions" was issued in February 2019, which described the expedited programs available to sponsors of regenerative medicine for serious conditions.

- Fast-track designation (FTD) could be granted to products demonstrating the potential to address an unmet medical need based on pre-clinical (such as an in vitro or animal model) or clinical data. Advantages of FTD include actions to facilitate development and expedite the review of the product, such as the possibility for a rolling review, as well as the eligibility of priority review.
- Breakthrough designation (BTD) could be granted to products that are intended to treat a serious condition and preliminary clinical evidence indicated a substantial improvement over available treatment. Products qualified as BTD could have all the benefits of FTD, in addition to intensive interaction with the FDA and organizational commitment involving senior managers, to help developers optimize their drug development program. A higher level of evidence is required for BTD than FTD since they are not only required to address the unmet clinical need but also the substantial advantages must be demonstrated. Products eligible for priority review designation must demonstrate significant safety or effectiveness improvements in the diagnosis, treatment, or prevention of serious conditions.
- Priority review implies that the FDA will accomplish the review of an MA application within 6 months compared to 10 months under standard review. Accelerated approval applies to products indicated for diseases with a long course requiring an extended period to measure clinical benefit. Under such circumstances, surrogate endpoints or intermediate clinical endpoints will be acceptable to evaluate the effectiveness of products, but confirmatory post-approval studies must be conducted to verify the products' effects on the definite clinical endpoint.

An expedited pathway tailored to regenerative medicine, called the Regenerative Medicine Advanced Therapy (RMAT) designation, was proposed in December 2016 at the call of the 21st Century Cures Act. Regenerative medicine eligible for RMAT designation should be intended to treat, modify, reverse, or cure a serious condition; in the meanwhile, preliminary clinical evidence indicates that the regenerative medicine therapy has the potential to address unmet medical needs for such a condition. Unlike BTD, no significant advantages over existing treatments are required for RMAT, while other factors would be considered to determine the eligibility of RMAT, such as the rigor of data collection, disease severity, and prevalence, as well as potential study bias. Advantages of the RMAT designation include all the benefits of FTD and BTD, including more intensive dialogues and interactions with regulators in an early stage of development to support developers to obtain accelerated approval and satisfy post-approval requirements (12).

1.1.2.3 Cell and Gene Therapies Approved in the USA

As of November 2022, a total of 16 cell and gene therapies were approved in the USA, without considering cord blood. Among them, most of them were indicated for oncology diseases (Table 1.3).

TABLE 1.3
Cell and Gene Therapies Approved in the United States

Generic Name	Trade Name	Indication	Disease Area	Approval Date
Laviv	Azficel-T	Indicated for improvement of the appearance of moderate to severe nasolabial fold wrinkles in adults.	Orthopedic therapy	20/06/2011
Autologous cultured chondrocytes on a porcine collagen membrane	Maci	Indicated for the repair of single or multiple symptomatic, full-thickness cartilage defects of the knee with or without bone involvement in adults.	Musculoskeletal disease	13/12/2016
Allogeneic cultured keratinocytes and fibroblasts in bovine collagen	Gintuit	Indicated for topical (non-submerged) application to a surgically created vascular wound bed in the treatment of mucogingival conditions in adults.	Vascular disease	09/03/2012
Talimogene laherparepvec	Imlygic	Indicated for the local treatment of unresectable cutaneous, subcutaneous, and nodal lesions in patients with melanoma recurrent after initial surgery.	Oncologic disease	27/10/2015
Tisagenlecleucel	Kymriah	Indicated for the treatment of patients up to 25 years of age with B-cell precursor acute lymphoblastic leukemia (ALL) that is refractory or in second or later relapse; adult patients with relapsed or refractory large B-cell lymphoma.	Oncologic disease	30/08/2017 (indication-AML) 13/04/2018 (indication extension-DLBCL)

(*Continued*)

TABLE 1.3 *(Continued)*
Cell and Gene Therapies Approved in the United States

Generic Name	Trade Name	Indication	Disease Area	Approval Date
Voretigene neparvovec-rzyl	Luxturna	Indicated for the treatment of patients with confirmed biallelic *RPE65* mutation-associated retinal dystrophy. Patients must have viable retinal cells as determined by the treating physician(s).	Ophthalmic disease	18/12/2017
Autologous cellular immunotherapy	Provenge	For the treatment of asymptomatic or minimally symptomatic metastatic castrate-resistant (hormone-refractory) prostate cancer.	Oncologic disease	29/04/2010
Axicabtagene ciloleucel	Yescarta	Indicated for the treatment of relapsed or refractory DLBCL, primary mediastinal large B-cell lymphoma, high-grade B-cell lymphoma, and DLBCL arising from follicular lymphoma.	Oncologic disease	18/10/2017
Onasemnogene abeparvovec-xioi	Zolgensma	Spinal muscular atrophy (SMA)	Neurologic disease	24/05/2019
Brexucabtagene autoleucel	Tecartus	Mantel cell lymphoma; acute lymphocytic leukemia	Oncologic disease	24/07/2020
Lisocabtagene maraleucel	Breyanzi	Diffuse large B-cell lymphoma; follicular lymphoma	Oncologic disease	05/02/2021
Idecabtagene vicleucel	Abecma	Multiple myeloma	Oncologic disease	26/03/2021
Allogeneic processed thymus tissue–agdc	Rethymic	For immune reconstitution in pediatric patients with congenital athymia.	Immune disease	08/10/2021

TABLE 1.3 *(Continued)*
Cell and Gene Therapies Approved in the United States

Generic Name	Trade Name	Indication	Disease Area	Approval Date
Allogeneic cultured keratinocytes and dermal fibroblasts in murine	Stratagraft	Adults with thermal burns containing intact dermal elements for which surgical intervention is clinically indicated (deep partial-thickness burns).	Dermal disease	15/06/2021
Ciltacabtagene autoleucel	Carvykti	Adult patients with relapsed or refractory multiple myeloma after four or more prior lines of therapy.	Oncologic disease	28/02/2022

1.1.3 JAPAN

1.1.3.1 New Released Pharmaceutical and Medical Device Act

Following the introduction of the Regenerative Medicine Promotion Act to promote the use of regenerative medicine in Japan, the Pharmaceutical and Medical Device Act (PMD Act), as the revised and updated version of Pharmaceutical Affairs Law, was enacted in November 2013 and enforced within a 1-year period. Regenerative medicine was defined as a new category of products separate from conventional pharmaceutical medicine and medical devices. According to the PMD Act, "regenerative medicine" was defined as products processed from the cells of a human or animal with intention to reconstruct, repair, or reform structures or functions of the human body (i.e., tissue-engineering products) or for the treatment or prevention of human diseases (i.e., cellular therapy products), and products intended for the treatment of disease in humans (or animals) transgened to express in human (or animal) cells (i.e., gene therapy products). Furthermore, the new Good Gene, Cellular and Tissue-based Products Manufacturing Practice (GCTP) was created under the revised PMD Act, which sought to set up appropriate regulation requirements in terms of manufacturing management and quality control of regenerative medicine. The new act enables industry or hospitals to produce regenerative medicines using blood derived from humans, which was forbidden in the former regulations.

1.1.3.2 Time-Limited and Conditional Market Authorization

One highlight proposed in the revised PMD Act was the introduction of a time-limited and conditional market authorization pathway. This allows regenerative medicines an accelerated market authorization by the Ministry of Health, Labor, and Welfare (MHLW) if the applicants could demonstrate likely efficacy and confirmed safety in preliminary clinical trials. Utilization of regenerative medicines received conditional market authorization limited to medical institutions with qualified experts and sufficient facilities. Regenerative medicines that received conditional approval were then automatically reimbursed by the Japanese health system, but required up to 30% co-payment from patients depending on the patient age and the type of disease condition. Following approval, manufacturers were obliged to carry out post-marketing studies to further collect efficacy and safety evidence for a maximum of 7 years. Afterwards, manufacturers should apply for a second evaluation to assess whether the actual performance of regenerative medicines meets the regulator's expectation and fulfills the requirements of full market authorization based on post-market evidence collected in the period of conditional market authorization. If regenerative medicines failed to achieve the previous claimed effectiveness or safety outcomes, the market authorization could be revoked by the MHLW. In September 2015, HeatSheet, a skeletal myoblast sheet product for the treatment of severe heart failure, became the first product granted conditional 5-year market approval in Japan based on the data collected from an open-label, uncontrolled, multicenter clinical trial of seven patients.

Furthermore, additional safety considerations were incorporated into new PMD Act, which stipulate that a patient registry for regenerative medicine must be put in place to carry out the post-marketing surveillance and require the regular report of the incidence of infectious disease. Furthermore, patients suffering from adverse events related to regenerative medicines were included under the umbrella of Relief Services for Adverse Health Effects, under which patients would get timely financial support in the case of severe side effects.

1.1.3.3   Act on the Safety of Regenerative Medicine

- Risk-based approach for reviewing the clinical research protocol

A second important law passed that aligned with the framework of the Regenerative Medicine Act was the Act on the Safety of Regenerative Medicine (RM Act), in which a regulatory framework was established to ensure the safety and ethicality of regenerative medicine in both the clinical research (not including the standard clinical trial complying with good clinical practice) and private medical practice (not covered by health insurance). One significant improvement of the RM Act was the implement of a risk-based approach when evaluating the appropriateness of clinical research. Regenerative medical techniques were classified into three categories based on their potential harm to human health: high risk (Class I), intermediate risk (Class II), and low risk (Class III). In general, somatic cells (such as the autologous differentiated cells for homologous use) with accumulated clinical experiences will fall into Class III, whereas human embryotic stem cells, iPS cells, and their derivatives will fall into Class I (5). Clinical research protocols for Class III techniques should be reviewed by the Certified Committee for Regenerative Medicine (CCRM), but there is no requirement to send the protocol to MHLW for approval. On the contrary, clinical research protocols for Class I and Class II techniques must be first reviewed by the Certified Special Committee for Regenerative Medicine (CSCRM) and then further approved by MHLW.

• SAKIGAKE Designation System

On June 2014, the "Strategy of SAKIGAKE" was announced by the MHLW, in which one major policy named the "SAKIGAKE Designation System" was proposed to promote the R&D of innovative medical products, including pharmaceuticals, medical devices, and regenerative medicine products. Products satisfying the specific criteria for SAKIGAKE designation will benefit from consistent prioritized consultation, pre-application consultation, priority review (rolling submission), assigning a manager as a concierge, and the extension of the re-examination period. After three rounds of applications, nine regenerative medicines have been granted SAKIGAKE designation until April 2018.

1.1.3.4 Regenerative Medicine Approved in Japan

As of September 2021, a total of 14 regenerative medicines were approved in Japan (Table 1.4).

TABLE 1.4
Regenerative Medicines Approved in Japan

Product Name	Indication	Disease Area	Product Classification	MA Fate
Collategene (beperminogene perplasmid)	Critical limb ischemia	Peripheral arterial disease	Gene therapy	20/02/2019: conditional and time-limited (5 years) approval
Kymriah (chimeric antigen receptor T-cell)	CD19-positive relapsed/refractory ALL; DLBCL in adults	Oncologic disease	Human somatic cell–processed products	20/02/2019: standard approval
STR01 (autologous bone marrow–derived mesenchymal stem cell)	Spinal cord injury	Spinal cord injury	Human somatic stem cell–processed product	28/12/2018
JACE (human autologous epidermis–derived cell sheet)	Indication extension: giant congenital melanocytic nevus	Dermal disease	Human somatic cell–processed product	29/09/2016

(Continued)

TABLE 1.4 *(Continued)*
Regenerative Medicines Approved in Japan

Product Name	Indication	Disease Area	Product Classification	MA Fate
Temcell HS (human allogeneic bone marrow–derived mesenchymal stem cell)	Acute GVHD after allogeneic HSCT	Adjunctive treatment for HSCT	Human somatic stem cell–processed products	18/09/2015
HeartSheet (human autologous skeletal myoblast–derived cell sheet)	Serious heart failure caused by ischemic heart disease	Cardiovascular disease	Human somatic stem cell–processed products	18/09/2015
JACC (autologous cultured cartilage)	Relief of symptoms of traumatic cartilage defects and osteochondritis dissecans (excluding osteoarthritis) for knee joints	Musculoskeletal diseases	Tissue-engineered products (approved as medical device)	27/07/2012
JACE (human autologous epidermis–derived cell sheet)	Degree II and degree III burn wounds	Dermal disease	Human somatic cell–processed product (approved as medical device)	06/08/2007
Kymriah (chimeric antigen receptor T-cell)	CD19-positive relapsed/refractory ALL or relapsed/refractory DLBCL in adults	Oncologic disease	Human somatic cell–processed products	20/02/2019: standard approval
STR01 (autologous bone marrow–derived mesenchy-mal stem cell)	Spinal cord injury	Spinal cord injury	Human somatic stem cell–processed product	28/12/2018
JACE (human autologous epidermis–derived cell sheet)	Indication extension: giant congenital melanocytic nevus	Dermal disease	Human somatic cell–processed product	29/09/2016
Delytac (teserpaturev)	Malignant glioma	Oncologic disease	Gene therapy	11/06/2021
Ocural (human autologous oral mucosa-derived epithelial cell sheet)	Repair of corneal epithelium defects by transplanting the cell sheet onto the ocular surface of patients with limbal stem cell deficiency	Optical disease	Tissue-engineered products	11/06/2021

TABLE 1.4 *(Continued)*
Regenerative Medicines Approved in Japan

Product Name	Indication	Disease Area	Product Classification	MA Fate
Nepic (human (autologous) corneal limbus– derived corneal epithelial cell sheet)	Repair of corneal epithelium defects by transplanting the cell sheet onto the ocular surface of patients with limbal stem cell deficiency	Optical disease	Tissue-engineered products	19/03/2020

1.1.4 AUSTRALIA

1.1.4.1 Australian Regulatory Guidelines for Biologicals

On 27 July 2018, the updated version of the Australian Regulatory Guidelines for Biologicals (ARGB) was released by the Therapeutic Goods Administration (TGA). Some autologous human cells and tissues (HCTs) may be excluded from all TGA regulation if they meet all of the following criteria: 1) collected from a patient who is under the clinical care of a medical or dental practitioner, 2) manufactured and used by the practitioner with primary responsibility for clinical care OR by a person or persons under the professional supervision of that practitioner, 3) indicated for a single indication in a single clinical procedure, and 4) HCTs are minimally manipulated and for homologous use. Examples of possible biologicals exempted from regulations included bone grafts for dental procedures and platelet-rich plasma.

1.1.4.2 Classification Before Inclusion in the ARTG

Under the ARGB regulatory framework, biologicals, except for those exempted from TGA regulation or supplied under authorization of unapproved use, must be included on the Australian Register of Therapeutic Goods (ARTG) to ensure the lawful supply of this product in Australia (Figure 1.1). Before biologicals can be included in the ARTG, they have to be classified according to the risk level associated with their use in patients. Biologicals classified into Class 1 and Class 4 must be mentioned in the Schedule 16 of the Therapeutic Goods Regulations (TGR) 1990, while biologicals classified into Class 2 and Class 3 could be evaluated based on two considerations: level of processing applied and intended use of the product.

- Class 1 biologicals were considered products with low risk and had an appropriate level of external governance and clinical oversight; therefore, they would be subject to less strict regulation. At the time of writing this manuscript, there is currently no product that has been defined as a Class 1 biological.

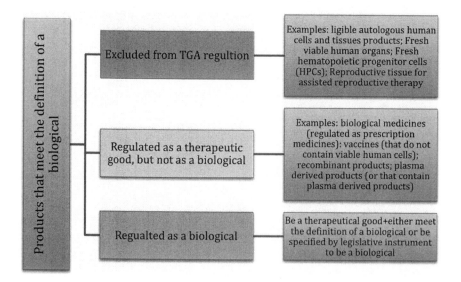

FIGURE 1.1   The regulation of regenerative medicines in Australia.

Source: Department of Health. Therapeutic Goods Administration

- Class 2 biologicals are considered low risk and are restricted to those that have been subject to only minimal manipulation and intended for homologous use.
- Class 3 biologicals are considered medium risk, comprising the biologicals that are subject to more than minimal manipulation or intended for non-homologous use (regardless of preparation method).
- Class 4 biologicals are considered high risk and are defined in Schedule 16 as biologicals containing live animal cell/issue/organs, or human cells that have been genetically modified to artificially introduce a non-basic function (such as a CAR T cell), or pluripotent stem cells along with their derivation. Sponsors should prepare and submit their application for inclusion on the ARTG to TGA. The Advisory Committee on Biologicals (ACB) will be consulted when necessary to seek independent and scientific advice in relation to the safety and efficacy of biological products. The total timeframe for dossier evaluation was about 200 to 220 working days, depending on the complexity of the dossiers, the timely response of the request, and advice from the ACB.

1.1.4.3   Cell and Gene Therapies Approved in Australia

As of March 2022, a total of six cell and gene therapies were approved in Australia (Table 1.5), including Imygic and Luxturna, which was regulated as a medicine in Australia, but classified as gene therapy in other countries.

TABLE 1.5
Cell and Gene Therapies Approved in Australia

Product Name	Indication	Disease Area	Product Type	MA Date
Luxturna	Confirmed biallelic RPE65 mutation–associated retinal dystrophy. Patients must have viable retinal cells as determined by the treating physician(s).	Optical disease	Medicine	13/08/2020
Tecartus (brexucabtagene autoleucel)	Mantel cell lymphoma; acute lymphocytic leukemia	Oncologic disease	Cell therapy, Class 4	21/07/2021
Yescarta (axicabtagene ciloleucel)	Relapsed or refractory DLBCL, primary mediastinal large B-cell lymphoma, high-grade B-cell lymphoma, and DLBCL arising from follicular lymphoma.	Oncologic disease	Cell therapy, Class 4	11/02/2020
Kymriah (tisagenlecleucel)	Relapsed/refractory B-cell acute lymphoblastic leukemia (ALL) in patients up to 25 years of age including children and relapsed/refractory diffuse large B-cell lymphoma (DLBCL) in adults.	Oncologic disease	Cell therapy, Class 4	19/12/2018
Chondrocytes-T (autologous cultured chondrocytes)	Cartilage lesions associated with the knee, patella, and ankle	Musculoskeletal diseases	Cell therapy, Class 3	26/05/2017
Imygic (talimogene laherparepvec)	Monotherapy for the treatment of melanoma in patients with unresectable cutaneous, subcutaneous, or nodal lesions after initial surgery	Oncologic disease	Medicine	21/12/2015

1.1.5 CANADA

1.1.5.1 Regulations on Biologicals

In Canada, biologicals, including cell and gene therapies, were regulated as Schedule D drugs by the Health Canada's Biologics and Genetic Therapies Directorate under the authority of the Food and Drug Act and its secondary

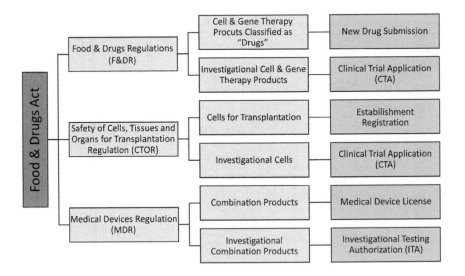

FIGURE 1.2   The regulation of regenerative medicines in Canada.

Source: Chisholm, et al. Current state of Health Canada regulation for cellular and gene therapy products: potential cures on the horizon. doi.org/10.1016/j.jcyt.2019.03.005. License number for reuse: 4970120095110

legislations. The most important regulation was the Food and Drug Regulations, in which regulatory requirements with respect to clinical trial design, pre-market approval, and market authorization were specified. Certain allogeneic cells and tissues that are minimally manipulated for homologous use are regulated under the Safety of Human Cells, Tissues and Organs for Transplantation Regulations. Some products may meet the definition of a medical device or combination products, and thus are regulated under the Medical Device Regulations or Health Canada's Policy on Drug/Medical Device Combination Products, respectively (Figure 1.2).

1.1.5.2   Strategies to Facilitate Market Authorization

An early market approval mechanism, called the Notice of Compliance with Conditions (NOC/c), was established to expedite the market authorization of promising medicines for the treatment of serious, life-threatening, or severely debilitating diseases wherein no therapy is presently marketed or a substantial improvement in the benefit/risk profile of the therapy is demonstrated. Sponsors must comply with all the post-market commitments agreed to under the conditional approval framework, including carrying out further confirmatory studies to verify the clinical benefits of the product in real-life practice, as well as conducting additional post-market surveillance activities to monitor safety. In 2012, Prochymal (remestemcel-L), used for acute graft-versus-host disease (aGvHD), was the first and the only cell therapy approved under the NOC/c. In addition, priority review of drug submissions is also established, which is a "fast-tracking" pathway for new therapies, preventatives, and diagnostic agents for serious, life-threatening, or severely debilitating diseases or conditions.

1.1.5.3 Cell and Gene Therapies Approved in Canada

As of March 2020, a total of seven cell and gene therapies were approved in Canada (Table 1.6), and four of them were CAR-T cell therapies indicated for hematological malignance.

TABLE 1.6
Cell and Gene Therapies Approved in Canada

Product Name	Indication	Disease Area	Product Classification	MA Date
Yescarta (axicabtagene ciloleucel)	CD19-directed genetically modified autologous T-cell immunotherapy indicated for adult patients with relapsed or refractory large B-cell lymphoma including DLBCL	Oncologic disease	Gene therapy	19/02/2019 (approved under NOC/c)
Kymriah (chimeric antigen receptor T-cell)	Relapsed/refractory B-cell acute lymphoblastic leukemia (ALL) in patients up to 25 years of age including children and relapsed/ refractory diffuse large B-cell lymphoma (DLBCL) in adults	Oncologic disease	Gene therapy	05/09/2018
Tellcell (also known as Prochymal, remestemcel-L)	Adult human mesenchymal stem cells (hMSCs) indicated for acute graft-versus-host disease (aGvHD)	Post-transplantation	Cell therapy	06/03/2014 (approved under (NOC/c))
Zolgensma (onasemnogene abeparvovec-xioi)	Spinal muscular atrophy (SMA)	Neurologic disease	Gene therapy	31/03/2021

(Continued)

TABLE 1.6 *(Continued)*
Cell and Gene Therapies Approved in Canada

Product Name	Indication	Disease Area	Product Classification	MA Date
Luxturna (voretigene neparvovec-rzyl)	Confirmed biallelic RPE65 mutation-associated retinal dystrophy	Optical disease	Gene therapy	20/07/2021
Abecma (idecabtagene vicleucel)	Multiple myeloma	Oncologic disease	Gene therapy	12/11/2021 (approved under (NOC/c))
Tecartus (brexucabtagene autoleucel)	Mantel cell lymphoma; acute lymphocytic leukemia	Oncologic disease	Gene therapy	11/01/2022

1.1.6 SOUTH KOREA

1.1.6.1 Before August 2020: Pharmaceutical Affairs Act

In South Korea, cell therapy and gene therapy, just like other biological products, are regulated by the Ministry of Food and Drug Safety (MFDS) under the authority of the Pharmaceutical Affairs Act (PAA). Additionally, there are several non-banding notifications and recommendations relevant to regulatory activity associated with cell therapy. The main divisions responsible for the regulation of cell therapy consist of the Biopharmaceuticals and Herbal Medicines Bureau (BHB) in the MFDS and the Biopharmaceuticals and Herbal Medicines Evaluation Department (BHED) in the National Institute of Food and Drug Safety (NIFDS).

- Risk-based and case-by-case approach

Although the unique and complex characteristic of cell therapy was considered, the overall principle of regulation on cell therapy was aligned with the regulation on conventional medicine. The regulation of cell therapy in Korea was featured as a risk-based and case-by-case approach, which was established by a scientific review with flexibility to balance risk and benefits. Just like conventional biological medicines, manufacturers of cell or gene therapies must ensure that the manufacturing process, pre-clinical trial, and clinical trial meet the corresponding GMP, GLP, and GCP standards. Evidence should be submitted to MFDS for evaluating whether the quality, efficacy, and safety of the applicant were sufficient for market authorization. A range of incentives were provided by the MFDS to encourage the development of cell therapies. A scientific advice program called the Majungmul (priming water) project (Figure 1.3) was available to provide tailored consultation and open communication with regard to market authorization applications of innovative biological products. A pre-review system allows a

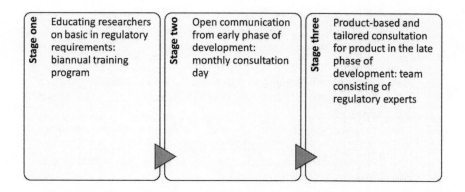

Stage one Educating researchers on basic in regulatory requirements: biannual training program

Stage two Open communication from early phase of development: monthly consultation day

Stage three Product-based and tailored consultation for product in the late phase of development: team consisting of regulatory experts

FIGURE 1.3 Majungmul (priming water) project in the South Korea.

sponsor to submit an IND or NDA document to MFDS section by section before submission of the full package. For a cell therapy that has an orphan drug designation, a confirmatory phase 3 study can be carried out after marketing authorization. Conditional approval could be granted for autologous keratinocytes and chondrocytes in Korea. It's worth noting that the gene therapy could only be approved in case of one of the following conditions: 1) intended for the treatment of genetic disease, cancer, AIDS, or other diseases that may be life-threatening or result in serious disorders; 2) an appropriate therapy is not available, or the safety and efficacy of the gene therapy product are obviously an improvement in comparison with other available therapies; or 3) other conditions deemed necessary for the prevention or treatment of disease by the commissioner of the KFDA.

- Post-market surveillance, re-examination, and re-evaluation

After market authorization, manufacturers are obliged to carry out post-market surveillance programs to track the actual performance of approved medicines. Periodic safety update reporting (PSDR) was required just like with conventional medicines. Re-examination of the safety and effectiveness of medicines in 3,000 patients for 6 years after market authorization was mandatory, and product labels should be re-evaluated periodically based on updated scientific information. In addition, a safety reporting system specific for every use of stem cell therapy, as well as long-term follow-up reporting for the patients enrolled in stem cell clinical trials, were required since July 2015 and December 2015, respectively.

1.1.6.2 After August 2020: Advanced Regenerative Medicine Act

The purpose of this new act is to establish a separate system for product lifecycle safety control and support from clinical research in advanced regenerative medicine (ARM) to commercialization of biopharmaceuticals, which aims to expand treatment opportunities for patients by facilitating clinical research in ARM and providing expedited procedures for biopharmaceuticals and to strengthen safety controls by implementing long-term follow-up requirements, etc.

The new legislation is primarily composed of three components: customized screening, priority screening, and conditional licensing.

- Customized screening is a pre-screening that allows developers to submit their permission and review data in advance according to their schedule. It is expected that the process will shorten the non-clinical and clinical period.
- Priority screening prioritizes advanced biopharmaceuticals over other drugs. While new drug screenings previously took about 115 days, the new priority screening process is expected to shorten the period by less than 100 days.
- Conditional licensing is designed to provide faster treatment opportunities for patients who do not have a treatment. It grants conditional marketing approval for companies that have completed phase 2 clinical trials for a specific drug while planning to conduct phase 3 clinical trials after receiving the sales approval.

1.1.6.3 Cell and Gene Therapies Approved in South Korea

As of May 2020, a total of 15 regenerative medicines were approved and marketed in South Korea (Table 1.7).

TABLE 1.7
Cell and Gene Therapies Approved in South Korea

Trade Name (active ingredient)	Indication	Therapeutic Areas	Product Type	MA Date
CureSkin (utologous dermal fibroblasts)	Depressed acne scar	Dermal disease	Cell therapy	11/05/2010
CreaVax-RCC (autologous dendritic cells)	Metastatic renal cell carcinoma	Oncologic disease	Cell therapy	15/05/2007
KeraHeal (autologous skin-derived keratinocytes)	Second-degree burns covering more than 30% of TBSA; third-degree burns covering more than 10% of TBSA	Dermal disease	Cell therapy	03/05/2006
KeraHeal-Allo (allogeneic skin-derived keratinocytes)	Deep second-degree burns	Dermal disease	Cell therapy	16/10/2015
Holoderm (autologous keratinocytes)	Deep second-degree burn; third-degree burn	Dermal disease	Cell therapy	10/12/2002

TABLE 1.7 *(Continued)*
Cell and Gene Therapies Approved in South Korea

Trade Name (active ingredient)	Indication	Therapeutic Areas	Product Type	MA Date
Kaloderm (autologous keratinocytes)	Deep second-degree burn; diabetic foot ulcer	Dermal disease	Cell therapy	21/05/2005 (second-degree burn) 24/06/2010 (diabetic foot ulcer)
Rosmir (autologous fibroblasts)	Improvement of nasojugal groove	Orthopedic disease	Cell therapy	28/12/2017
Cupistem (autologous adipose tissue–derived mesenchymal stem cell)	Crohn-related fistula	Gastrointestinal disease	Cell therapy	18/06/2012
Queencell (autologous adipose tissue–derived adipose cell)	Subcutaneous tissue defect	Dermal disease	Cell therapy	26/03/2010
Cellgram (autologous bone marrow–derived mesenchymal stem cell)	Improving heart functions and lowering the MACE through the improvement of left ventricular ejection fraction (LVEF) in acute myocardial infarction (AMI) patients	Cardiovascular disease	Cell therapy	01/07/2011
Neuronata-R (autologous bone marrow mesenchymal stem cell)	Amyotrophic lateral sclerosis	Neurologic disease	Cell therapy	30/06/2014
ChondronTM (cultured autologous chondrocytes)	Focal cartilage defect of the knee	Musculoskeletal diseases	Cell therapy	30/01/2001
RMS Ossron (cultured autologous osteoblasts)	Focal bone formation	Musculoskeletal diseases	Cell therapy	26/08/2009
Immuncell-LC (autologous activated T-cell)	Liver cancer (hepatocellular carcinoma)	Oncologic disease	Cell therapy	06/08/2007

1.1.7 China

1.1.7.1 Regulation of Biological Products

In realization of the rapid development in gene therapy, "Principles on Human Gene Therapy Research and Product Quality Control" was published by the State Food and Drug Administration of China (SFDA) in March 2003. This document outlined the basic requirements regarding gene therapy in clinical trials, the manufacturing process, quality control, and efficacy and safety tests. The general principle was quite encouraging to promote the research of gene therapy, and more flexibility on the quality control of innovative therapy techniques was allowed considering the unique characteristics of gene therapy compared to conventional chemistry products. In 2009, the China Ministry of Health classified cellular therapy and gene therapy as category 3 medical technologies, which refers to technologies considered to be associated with the highest risk, with complex ethical issues, and have not yet been scientifically proven in terms of therapeutic value. Under such regulation, approval from a technical audit board is required before a cellular therapy product can be used. In 2015, the National Health and Family Planning Commission of the People's Republic of China (NHFPC) announced that the requirement of approval before use of category 3 medical technology (including gene and cellular therapy) was abolished, and the hospitals are the responsible entity for quality control as well as risk management of category 3 medical technology (including gene and cellular therapy). Stem cell–based products were regulated as biological drugs rather than medical technology after the publication of "Regulation for Stem Cell Clinical Trials" by NHFPC in August 2015. This regulation stipulated that stem cell clinical trials could only be conducted in tier 3A hospitals, and patients participating in stem cell clinical trials must not be subject to any fees. The Experts Committee and Ethics Review Committee at the provincial level were required to establish regulations on clinical stem cell research in their jurisdiction. In 2017, the "Guideline on the Research and Review Process of Cellular Product" was issued by the China Food and Drug Administration (CFDA), which could be deemed the primary regulatory document of cellular products in China. A range of more flexible policies were introduced in this guideline, for instance, the acceptance of data obtained from non-registered clinical trials for market authorization approval, allowing the sponsors to self-design the phases of clinical trials by considering the unique characteristic of the cell therapy instead of strict phase 1 to 2 clinical trials for conventional products, and the exemption of a pharmacokinetics study for products that are infeasible to conduct. Additionally, market authorization applications for cellular and gene therapies have a high possibility to be evaluated under the "priority review and approval" process because they are featured with advanced manufacturing techniques, innovative therapy methods, and significant clinical advantages.

1.1.7.2 Cell and Gene Therapies Approved in China

As of March 2022, four gene therapies were approved in China (Table 1.8).

TABLE 1.8

Cell and Gene Therapies Approved in China

Product Name	Indication	Disease Area	Product Classification	MA Date
Gendicine (wild-type p53 gene)	Head and neck squamous cancers	Oncologic disease	Gene therapy	October 2003
Oncorine (recombinant human adenovirus type 5 injection)	Head and neck squamous cancers	Oncologic disease	Gene therapy	September 2006
Yescarta (axicabtagene ciloleucel)	Adult patients with relapsed or refractory large B-cell lymphoma	Oncologic disease	Gene therapy	June 2021
Carteyva (relmacabtagene autoleucel)	Adult patients with relapsed or refractory large B-cell lymphoma (r/r LBCL)	Oncologic disease	Gene therapy	September 2021

1.1.8 SINGAPORE

1.1.8.1 Regulation of Cell, Tissue, and Gene Therapy Products

Cell, tissue, and gene therapy products (CTGTPs) were regulated as medicinal products under the Medicines Act (1975). The Ministry of Health (MOH) will regulate the clinical use of CTGTPs products, while the Health Sciences Authority (HSA) will regulate the quality, safety, and efficacy of such products. CTGTPs could be differentiated as low-risk or high-risk products based on the degree of processing, the function they serve in the recipient, and the combination with other product types. The risk classification of CTGTPs will be conducted on a case-by-case basis at the pre-submission meeting between HSA and the applicant. High-risk CTGTPs refer to products that have been subject to substantial manipulation; are intended for non-homologous use; or are combined or used in conjunction with a drug, biologic, or device. Only high-risk CTGTP products are regulated as medicinal products and are subject to appropriate standards of quality, safety, and efficacy like other medicinal products, such as obtaining a Clinical Trial Certificate (CTC) for high-risk CTGTPs issued by HSA before the initiation of a clinical trial, GMP compliance for manufacturing facilities, and fulfillment of post-market obligations. An Innovative Office was set up in April 2018 to support the development of biomedical (including advanced therapies such as CTGTP products) sectors by providing scientific and regulatory advice from non-clinical development strategies to regulatory dossier submission.

TABLE 1.9
Cell and Gene Therapies Approved in Singapore

Product Name	Indication	Disease Area	Product Classification	MA Date
Kymriah (chimeric antigen receptor T-cell)	Relapsed/refractory B-cell acute lymphoblastic leukemia (ALL) in patients up to 25 years of age including children and relapsed/refractory diffuse large B-cell lymphoma (DLBCL) in adults.	Oncologic disease	Gene therapy, class 2	03/03/2021
Luxturna (voretigene neparvovec-rzyl)	Confirmed biallelic RPE65 mutation-associated retinal dystrophy. Patients must have viable retinal cells as determined by the treating physician(s).	Optical disease	Gene therapy, class 2	20/01/2022

The specific regulation for cell and gene therapies, named "Health Products (Cell, Tissue and Gene Therapy Products) Regulations 2021," came into operation on 1 March 2021. CTGTPs are risk-stratified into two classes:

- Class 1 (lower risk): CTGTP that satisfies ALL of the following criteria:
  - Minimally manipulated, i.e. biological characteristics or functions of the cell or the structural properties of the tissue are not altered
  - Intended for homologous use (performing the same function and administered at the same anatomical site or histological environment in the recipient as in the donor)
  - Not combined or used in conjunction with therapeutic products or medical devices
- Class 2 (higher risk): Other CTGTPs that are not classified as a Class 1 CTGTP.

1.1.8.2   Gene Therapies Approved in Singapore

As of March 2022, two gene therapies were approved in Singapore (Table 1.9).

1.1.9   INDIA

1.1.9.1   Regulation of Cell, Tissue, and Gene Therapy Products

The Central Drug Standards Control Organization (CDSCO) is the regulatory authority responsible for the drug (including cell and gene therapy products) administration activity in India. The Cell Biology Based Therapeutic Drug

Evaluation Committee (CBBTDEC) within the Biological Division in CDSCO was constituted in 2010 for the internal evaluation of all proposals in matters of clinical trials and market authorization regarding cell therapy. Further, the Apex Committee of CDSCO recommended that the Investigational New Drug proposals evaluated by CBBTDEC shall be directly placed before the Apex Committee without going through the Technical Committee, as was required earlier. Another milestone has been the provision of conditional approval of cell-based products for unmet needs if the product is demonstrating substantial safety and efficacy trends. The updated "National Guidelines for Stem Cell Research" was jointly issued by the Indian Council of Medical Research and the Department of Biotechnology in 2017 (13). Stem cell research could be defined as one of three categories based on the source of stem cells and the nature of the experiments, which are permissible, restricted, and prohibited research. Major highlights introduced in the latest guideline include the mandatory registration of the Institutional Committee for Stem Cell Research (ICSCR) and Institutional Ethics Committee (IEC) with the National Apex Committee for Stem Cell Research and Therapy (NAC-SCRT) and CDSCO, respectively, and undertaking clinical trials only at institutes with registered IC-SCR, IEC, and Good Manufacturing Practice (GMP) and Good Laboratory Practice (GLP) certified facilities. In April 2018, the draft Drugs and Cosmetics (Amendment) Rules of 2018 first proposed to regulate stem cell– and cell-based products processed by means of substantial or more than minimal manipulation as drugs. This proposed rule would apply to both autologous stem cells from a patient's own body and homologous stem cells from another human. This means all clinical trials using stem cells and other cell-based products must get permission from the Drug Controller General of India (GCGI) before clinical trials can be started; clinical use of stem cells (for treatment) and their products can only be allowed when the safety and efficacy have been proved by clinical trials, as well as manufacture and sale licenses must be obtained for the lawful use of stem cell– and cell-based products in India.

1.1.9.2   Cell and Gene Therapies Approved in India

As of April 2020, four cell or gene therapies were approved in India (Table 1.10).

TABLE 1.10

Cell and Gene Therapies Approved in Singapore

Product Name	Indication	Disease Area	Product Classification	MA Date
ChondronTM (Chondron ACI)	Cartilage defects of the joints	Musculoskeletal diseases	Cell therapy	April 2017
Ossgrow (Ossron ABI)	Avascular necrosis of the hip	Musculoskeletal diseases	Cell therapy	April 2017

(Continued)

TABLE 1.10 *(Continued)*
Cell and Gene Therapies Approved in Singapore

Product Name	Indication	Disease Area	Product Classification	MA Date
Apceden (autologous monocyte-derived mature dendritic cells)	Four cancer indications: prostate, ovarian, colorectal, and non–small cell lung carcinoma	Oncologic disease	Cell therapy	March 2017
Stempeucel (expanded adult human bone– marrow derived pooled allogeneic mesenchymal stromal cells)	Critical limb ischemia due to Buerger disease	Peripheral disease	Cell therapy	May 2016

1.1.10 NEW ZEALAND

All human medicines (including gene therapy and other protocols involving the administration of nucleic acids) are regulated under the Medicine Act (1981). However, the legislative and regulatory regimes governing research with cell, gene, and tissue-based products in New Zealand are complex, which mainly include the "Human Tissue Act (1964)," "New Zealand Public Health and Disability Act," "Human Assisted Reproductive Technology Act" (2004), and "Hazardous Substances and New Organisms (HSNO) Act." These legislations constitute the regulatory framework for different aspects relevant to research involving cell- or gene-based therapies, such as the biological sample collection, research design, establishment of a public database, ethical considerations, and environmental risk management. "Guidelines for Using Cells from Established Human Embryonic Stem Cell Lines for Research" was published in 2006, which imposed a number of restrictions on stem cell research, such as a mandatory ethical review and approval of human embryonic stem cell (hESC) application by an appropriate ethics committee prior to the commencement of research. A scientific assessment for the approval of clinical trials involving the introduction of nucleic acids or cells into human subjects is undertaken by the Gene Technology Advisory Committee (GTAC) under the Health Research Council (HRC) of New Zealand. Afterwards, GTAC will provide the director-general of health with their recommendation as to whether the trial would be approved. Assessment of new medicines (including cell and gene therapy) applications is under the responsibility of the New Zealand Medicines and Medical Devices Safety Authority (MEDSAFE). When MEDSAFE is not able to make the recommendation for the consent grant, for example, in the case of the application concerning a novel technology (such as gene therapy), the application will be referred to the Medicines Assessment Advisory Committee (MACC) for further advice. After the evaluation, MEDSAFE or MACC make a recommendation to the minister of health, who will decide whether to approve the medicine.

1.2   COMPARISONS BETWEEN COUNTRIES

1.2.1   INTERNATIONAL CLASSIFICATION AND DEFINITIONS OF RMs

The EU, United States, Japan, and South Korea have standard, formal definitions for RMs, albeit different terminologies for RMs are used.

- Japan and South Korea define and classify RMs into three product categories, including cell therapy medicinal products (CTMPs), gene therapy medicinal products (GTMPs), and tissue-engineered products (TEPs).
- The EU and United States use four categories, including CTMP, GTMP, TEP, and combined/combination products.
- In the United States, RMs are referred to as "regenerative medicine advanced therapies" (RMATs), which may be regulated as drugs, biological products, or medical devices.
- In the EU, RMs are defined as "advanced therapy medicinal products" (ATMPs) under the European Commission (EC) regulation No. 1394/2007 and are regulated as drugs.
- In Japan, the term "regenerative medicine" is used in the Japan Pharmaceutical and Medical Device Act (PMD Act) and is regulated under a unique pathway separate from drugs or biological products.
- In South Korea, the concept of "advanced regenerative medicine" has been introduced in the newly launched Advanced Regenerative Bio Act.

In Australia, Canada, Singapore, New Zealand, China, and India, no standard, formal definitions for RMs are available. RMs are generally regulated and defined as biological products in Australia, Canada, Singapore, and China, but are regulated as drugs in New Zealand and India. Differences were observed in the terminology used and product categories qualifying as RMs. In Australia, Singapore, and New Zealand, RMs comprised three product categories: cell therapy, gene therapy, and tissue-engineered products. In Canada, China, and India, cell therapy and gene therapy are classified as RMs, while no firm conclusions regarding whether tissue-engineered products are classified as either RMs or medical devices could be made.

In Australia, cell-based products, tissue-based products, and immunotherapy products containing human cells (e.g., chimeric antigen receptor [CAR] T-cell therapy) are included in the definition of "biologicals." Genetically modified products are regulated as biological medicines. In Singapore, all three product categories, cell products, gene products, and tissue-based products, are all regulated as "biological medicinal products." Canada defined "biological drugs

(Schedule D)" as originating from living organisms or from their cells and often manufactured using advanced biotechnology processes. It may be observed that gene therapies and many cell therapies will be captured and regulated as "biological drugs." However, it will be problematic to decide whether tissue-engineered products should be classified as biological products, medical devices, or combined products. In China, gene products and somatic cell products are regulated as "therapeutic biological products." Tissue-engineered products with viable cells are regulated as combined products. In New Zealand, gene therapies and/or other biotechnologies (such as cell- and tissue-based products) that are administered to human beings for therapeutic purposes are regulated as "medicine." Similarly, in India, stem cell–derived products and gene therapeutic products were defined and regulated as drugs.

1.2.2   RM Legislation and Frameworks

Among the nine investigated countries and the EU, four countries and the EU have established specific legislation for RMs:

- EC No. 1394/2007 put in place in 2007.
- AGRB implemented in May 2011 and updated in July 2018.
- Japan Regenerative Medicine Promotion Act released in 2013.
- South Korea's "Act on Advanced Regenerative Medicine and Advanced Biopharmaceuticals," will take effect in August 2020.
- In the United States, although there was no specific RM legislation available, a comprehensive RM policy framework was announced by the Food and Drug Administration (FDA) in November 2017.
- In Singapore, Health Products (Cell, Tissue, and Gene Therapy Products) Regulations 2021 came into operation on 1 March 2021.

In the remaining four countries (Canada, New Zealand, China, and India), RMs were regulated as conventional drugs or biological products under the oversight of primary pharmaceutical acts and/or law. However, guidelines, recommendations, or notifications specific to RMs were available to stipulate the tailored requirements regarding basic research, manufacturer practices, quality control, clinical trials, review processes, and MA procedures. Additionally, the US policy framework for RMs outlined a suite of guidelines, such as guidelines for "expedited programs for RMs for serious conditions." The FDA has also released several disease-specific (draft) guidelines for RMs, such as guidelines for gene therapies in rare diseases.

1.2.3   International RM Market Authorization Procedures

- **RMs exempt from MA regulation**

Six countries (USA, Australia, South Korea, Japan, Singapore, and India) and the EU have adopted risk-based regulatory approaches, which stipulate that eligible

RMs that were minimally manipulated for homologous use would be exempted from MA regulation.

Apart from the requirements for processing level and intended use, there were additional requirements for achieving regulatory exemption in the EU, the United States, and Australia. In the EU, eligible ATMPs must satisfy all the criteria of the "hospital exemption (HE)" rules:

- Be prepared on a non-routine basis according to specific quality standards
- Be utilized within the same member state in a hospital under the exclusive professional responsibility of a medical practitioner
- Comply with an individual medical prescription for a custom-made product for an individual patient

The FDA requires that exempted products (so-called "Section 361 HCT/Ps" of the Public Health Service [PHS] Act) must be non-combined, free of systematic effect, and independent upon the metabolic activity of living cells for its primary functions.

The TGA in Australia emphasized that autologous human cell and tissue products qualified for regulatory exemption must be manufactured and administered by the practitioners bearing the primary responsibility for clinical care, and they must target a single indication in a single clinical procedure.

Health Canada claimed that all autologous cell therapies, except for minimally manipulated lymphohematopoietic cells for homologous use, were new drugs and must be licensed for sale through the standard MA process. No similar regulation exemption exists in Canada.

- **Current RM MA procedures**

Only in Japan do RMs have a unique regulatory pathway: a conditional, time-limited MA pathway established under the PMD Act. RMs may be eligible for conditional MA for a maximum of 7 years, if the applicants demonstrate likely efficacy and confirm safety in preliminary clinical trials. Afterwards, holders of the conditional MA must submit a second application to the Pharmaceuticals and Medical Devices Agency (PMDA) to determine whether the RMs fulfill the requirements for standard MA based on post-market evidence.

In South Korea, a new regulation for advanced RMs will be enforced in August 2020, in which a similar pathway as conditional approval in Japan is proposed. It stipulates that conditional MA may be granted based on phase 2 clinical trials proving the clinical effectiveness, with the condition that phase 3 studies are conducted after commercialization.

In the EU, RMs go through a similar assessment process as other medicines under the centralized procedure by the Committee for Medicinal Products for Human Use (CHMP), with an additional step involving the Committee for Advanced Therapies (CAT), which drafts the initial opinions before CHMP delivers a final recommendation on the MA decision. Similarly, in the FDA, the Cellular, Tissue and Gene Therapies Advisory Committee evaluates the submitted evidence of an RM before the Center for Biologics Evaluation and Research (CBER) makes the ultimate decision.

In Australia, all RMs are classified in one of four classes: Class 1 and Class 2 are categorized as low risk, Class 3 as medium risk, and Class 4 as high risk, prior to inclusion on the Australian Register of Therapeutic Goods (ARTG) according to their processing level and intended use. Each category is subject to differing levels of requirements for supporting efficacy and safety evidence.

In New Zealand, the MA dossier evaluation is transferred to the Medicines Assessment Advisory Committee (MAAC) in the case of gene therapy applications. The MAAC will recommend to the minister's delegate whether MA should be granted or not. In Singapore, RMs re classified into either Class 1 lower-risk products or Class 2 higher-risk products.

In the remaining countries investigated, Canada, India, and China, RMs undergo the same MA procedures as conventional pharmaceuticals.

1.2.4   REGULATORY ASSESSMENT CONSIDERATIONS FOR RMs

Given the technical novelty, complexity, and specialty of RMs, more conditions and considerations in the evidence assessment of the MA application submissions have been considered by the FDA in the United States, the European Medicines Agency (EMA) in the EU, the TGA in Australia, Health Canada in Canada, the Ministry of Food and Drug Safety (MFDS) in South Korea, and the National Medical Products Administration (NMPA) in China.

- The FDA has suggested that novel study designs for early-phase clinical trials for RMs are necessary to address their uniqueness. Compared to the clinical trials for traditional medicines, some modifications in RM clinical trials regarding the patient eligibility criteria, starting dose choice, efficacy endpoint, and statistical methods are allowed.
- The EMA has stated that technical requirements, in particular, the type and the degree in quality of pre-clinical and clinical data, necessary to demonstrate the quality, safety, and efficacy, would be tailored to RMs. The EMA has differentiated the clinical trials for RMs as exploratory and pivotal clinical trials, rather than the traditional clinical trial phases.
- The Australia TGA acknowledged that adaptive trial designs and other cross-phase studies, such as phase 2b-2, might be more appropriate for the development of certain therapeutic goods, including RMs.
- Health Canada has indicated that dose estimations for RMs should be based on previous clinical experience with similar cell types when possible and that valid surrogate endpoints may be acceptable for RMs, especially for products indicated for rare diseases.
- The MFDS has acknowledged that not all the MA requirements for biological products would be applicable to gene therapy products and cell

therapy products. In the case of autologous cell therapy products, data submission of bridging studies from clinical study data could be waived.

- The NMPA has agreed with the use of non-registered clinical trial data for MA applications on the condition that the credibility, integrity, accuracy, and traceability of non-registered clinical data were justified. Clinical trials should be designed according to the characteristics of applicants, which could generally be classified as early exploratory and confirmatory studies.

1.2.5 ACCEPTANCE OF ALTERNATIVE EVIDENCE GENERATION

Considering the challenges of demonstrating the efficacy and safety of RMs in traditional RCTs, several countries, the United States, Japan, Canada, Singapore, and China, as well as the EU have expressed readiness to accept clinical evidence generated outside of traditional RCTs.

- In "Modern Trial Design and Evidence Development" of the 21st Century Cures Act, the FDA declared it would consider real-world evidence (RWE), such as observational data, patient registries, or data generated from other sources informing the health status, to support regulatory decisions. Moreover, in the case of gene therapies for rare diseases, natural history data may be accepted as a historical control group if it was adequately matched with the treatment group.
- In the EMA, although no specific documents or guidelines for RWE exists, RWE has long been used for accelerated assessment and post-authorization risk management. The "Adaptive Pathways," implemented to enable faster MA to new medicines, have been built on the collection of RWE through patient registries and other pharmacovigilance tools. However, Adaptive Pathway was criticized for its reliance on RWE to compensate for the weak evidence of drug efficacy and safety at the time of MA, which could be problematic. RWE may reveal incorrect conclusions about the benefit-risk balance of the innovative drugs, exposing patients to products with uncertain effectiveness and safety (12).
- In the PMDA, new trends on the utilization of RWE for regulatory purposes have emerged, with new consultations on the appropriateness of using patient registry data in MA application for drugs and RMs started in April 2019.
- In Health Canada, the project "Strengthening the use of real world evidence for drugs" was initiated in August 2018 and was aimed to optimize the use of RWE for the assessment and monitoring of drug efficacy and safety throughout the whole drug lifecycle, such as in the case of rare diseases or ethical issues in conducting traditional RCTs.
- In the NMPA of China, the "Guideline for the Use of Real-World Evidence for Research and Regulatory Review" was launched in January 2020. For example, pragmatic clinical trials or the use of RWE as a control

group for a single-arm study could be accepted to support regulatory decision-making, but intensive communication with the NMPA to discuss the feasibility and methodology of collecting RWE is required. In all jurisdictions, allowing the use of non-RCT data for MA application was not exclusively applicable to RMs, but to all innovative drugs with substantial therapeutic benefits.

No relevant regulations, guidelines, and/or documents to support the use of RWE in MA processes were available in Australia, New Zealand, South Korea, and India.

1.2.6 EXPEDITED MA ASSESSMENT PROGRAMS FOR RMS

All nine investigated countries and the EU have established expedited programs to facilitate the review and assessment of innovative therapies, although most of them are not exclusively applicable to RMs.

Multiple types of drug designations may be granted to products indicated for severe conditions with unmet clinical needs, including rare pediatric disease designation, fast-track designation, breakthrough designation, and RMAT by the FDA; PRIME designation by the EMA; and SAKIGAKE designation, as well as RM-specific orphan drug designation by the PMDA.

Additionally, priority review in the United States, Japan, Australia, Canada, Singapore, South Korea, and China; accelerated assessment in the EU; and priority assessment in New Zealand shared the common principle that the review timeframe for applicants may be reduced compared to the timeframe under the standard review procedure. Abbreviated evaluation in New Zealand and abridged evaluation in Singapore are intended to be a simpler and quicker evaluation process than the standard evaluation process, in which the review of overseas regulatory evaluation reports form the basis of the evaluation (Table 1.11).

TABLE 1.11
Expedited Approval Program in Each Region

Regions[a]	Expedited Approval Programs
European Union	• Conditional approval
	• Approval under exceptional circumstances
	• Adaptive pathway (suspended)
	• Priority medicine (PRIME)
	• Accelerated assessment
United States	• Fast-track
	• Breakthrough designation
	• Priority review
	• Accelerated approval
	• Rare pediatric disease designation
	• RMAT designation

TABLE 1.11 *(Continued)*
Expedited Approval Program in Each Region

Regions[a]	Expedited Approval Programs
Japan	• Conditional, time-limited market authorization • Priority review • SAKIGAKE designation • Regenerative medicine–specific orphan drug designation
Australia	• Provisional approval • Priority review
Canada	• Access to drugs in exceptional circumstance • Notice of compliance with conditions (NOC/c) • Priority review
New Zealand	• Abbreviated evaluation • Priority assessment • Provisional consent
South Korea	• Conditional approval • Fast-track review • Priority review
China	• Conditional approval • Priority review and approval • Designation for breakthrough therapeutics • Designation for overseas new drugs with high unmet clinical needs
Singapore	• Abridged evaluation • Priority review
India	• Conditional approval • Fast-track approach

[a]: Except for the RMAT designation in the United States and regenerative medicine–specific orphan drug designation in Japan, all the other expedited programs were not exclusively applicable to regenerative medicines. However, regenerative medicines addressing high unmet clinical needs have a high possibility to benefit from the previously mentioned expedited programs.

1.2.7 ACCELERATED APPROVAL PROGRAMS FOR RMs

All investigated countries and the EU, except for Singapore, have established accelerated approval pathways for innovative therapies where the benefit of immediate availability outweighs the risk of less comprehensive data. These pathways included conditional MA (EU, Japan, South Korea, China, and India), NOC/c (Canada), accelerated approval (USA), provisional approval (Australia), and provisional consent (New Zealand). The EU stated that the conditional MA can only be valid for a 1-year period. Conditional approval in New Zealand, Australia, and Japan is limited to a maximum of 2 years, 6 years, and 7 years, respectively. In all countries and the EU, standard MA could be granted when the evidence deficiency in the initial MA application has been addressed. Otherwise, conditional MA could be revoked if the products fail to confirm

efficacy and safety in post-market studies. Approval under exceptional circumstances in the EU acknowledges that the comprehensive data are impossible
to collect for certain products even after MA, possibly due to rare diseases or
unethical considerations; thus it normally will not convert to standard MA in
the EU (Table 1.12; Table 1.13).

TABLE 1.12
Accelerated Market Authorization Programs in Europe, USA, and Japan

Country	Accelerated Program	Eligible Criteria	Advantages
Europe	Conditional market authorization	• Positive risk-benefit balance • Unmet clinical needs will be fulfilled • It is likely to provide comprehensive data after MA • Benefit of immediate availability outweighs the risk of lees comprehensive data	• Accelerated approval on condition that scientific obligation must be fulfilled to collect confirmatory data • Converted into standard market authorization once comprehensive data were collected
	Approval under exceptional circumstance (EC)	Applicant is unable to provide comprehensive data on the efficacy and safety under normal conditions of use due to 1) disease rarity or 2) ethical reasons	
	Accelerated assessment	• Expected to be of major public health interest • Particularly from therapeutic innovation perspective	Reduce CHMP review timeframe to 150 days compared to 210 days under standard review
	Priority medicine (PRIME) designation	• Offer a major therapeutic advantage over existing treatments • Benefit patients without treatment options • Show its potential to benefit patients with unmet medical needs based on early clinical data	• Appoint a rapporteur from CHMP or CAT • Intensive guidelines on the overall development plan and regulatory strategies • Scientific advice at key development milestones, involving additional stakeholders, such as HTA body • Potential for accelerated assessment

TABLE 1.12 *(Continued)*
Accelerated Market Authorization Programs in Europe, USA, and Japan

Country	Accelerated Program	Eligible Criteria	Advantages
United State	Fast-track	• Indicated for serious conditions • Fill an unmet medical need defined as providing a therapy where none exists or providing a therapy which may be potentially better than available therapy (such as superior effectiveness or avoiding serious side effects)	• More frequent meetings with FDA to discuss the drug's development plan and data collection • More frequent written communication from FDA to discuss the clinical trials design and use of biomarkers • Rolling review
	Accelerated approval	• Drugs for serious conditions that filled an unmet medical need	Allow faster drug approval based on 1) a surrogate (such as a laboratory measurement, radiographic image) or 2) an intermediate clinical endpoint (a measure of a therapeutic effect that is considered reasonably likely to predict the clinical benefit)
	Priority review	• Drugs that have significant improvements in the following aspects: 1) increased effectiveness, 2) eliminate or significantly reduce the drug's adverse effects, 3) enhancement in patient compliance, and 4) better safety or effectiveness in a new subpopulation	Reduce the FDA review time to 6 months compared to 10 months under standard review
	Breakthrough designation	• Intend to treat a serious condition • Preliminary clinical evidence indicates that the drug may demonstrate substantial improvement over available therapy on a clinically significant endpoint	• All fast-track designation features • Intensive guidance on an efficient drug development program, beginning as early as phase 1 • Organizational commitment involving senior managers

(Continued)

TABLE 1.12 *(Continued)*
Accelerated Market Authorization Programs in Europe, USA, and Japan

Country	Accelerated Program	Eligible Criteria	Advantages
	Regenerative medicine advance therapy (RMAT) designation	• Regenerative medicines that are intended to treat, modify, reverse, or cure a serious or life-threatening disease or condition • Preliminary clinical evidence indicates that the drug has the potential to address unmet medical needs for such disease or condition	• Intensive guidelines on drug development as early as phase 1 • Early interaction to discuss potential surrogate or intermediate endpoints • Organizational commitment involving senior managers • Statute addresses potential ways to support accelerated approval and satisfy post-approval requirements
Japan	Time-limited, conditional market authorization	• Regenerative medicines showing likely efficacy and confirmed safety in early clinical studies	• A maximum of 7 years of market authorization on the condition that further data will be collected during the conditional MA period
	SAKIGAKE designation	• Product innovativeness • Target disease condition should be serious or life-threatening, or have no available curative treatments • Significantly improvement in effectiveness or safety compared to existing treatments • Develop the product rapidly and file an application for approval in Japan, ahead of other countries	• Consistent prioritized consultation • Pre-application consultation • Prioritized review aiming for a further reduction in the total review period to 6 months compared to 9 months in ordinal priority review and 12 months in standard review • Assigning a PDMA manager as a concierge • Extension of re-examination period • Potential of 10-20% premium at drug price

TABLE 1.12 *(Continued)*
Accelerated Market Authorization Programs in Europe, USA, and Japan

Country	Accelerated Program	Eligible Criteria	Advantages
	Regenerative medicine–specific orphan drug designation	• Prevalence of the disease covered by the indication of the product concerned is less than 50,000 patients in Japan • Indicated for serious disease with high medical needs	• Subsidy granting direct expenses for the development and authorization of such products • PMDA provides advice and consultation concerning the interpretation of designation criteria and other regulatory matters • Tax credits for the direct expense required during the subsidy period of research and development stage
	Priority review	• Target a serious or life-threatening condition • Demonstrate clinical advantages over existing therapies in terms of safety, efficacy, or patient quality of life	• PDMA review time reduced to 9 months compared to 12 months under standard review

TABLE 1.13
Regenerative Medicines Approved via Expedited Regulatory Programs

Country[a]	Drugs Approved Under Accelerated Pathways	Drugs With ODD[b]
USA	• Fast track (4): Hemacord, Imlygic, Provenge, Zolgensma • Priority review (7): Kymriah, Luxturna, Yescarta, Zolgensma, Tecartus, Stratagraft, Carvykti • Accelerated approval (2): Carticel, Tecartus, • Breakthrough designation (9): Kymriah, Luxturna, Yescarta, Zolgensma, Tecartus, Breyanzi, Abecma, Rethymic, Carvykti	(11) Imlygic, Kymriah, Luxturna, Yescarta, Zolgensma, Tecartus, Breyanzi, Abecma, Rethymic, Stratagraft, Carvykti

(Continued)

TABLE 1.13 *(Continued)*
Regenerative Medicines Approved via Expedited Regulatory Programs

Country[a]	Drugs Approved Under Accelerated Pathways	Drugs With ODD[b]
	• Rare pediatric disease designation (4): Kymriah, Luxturna®, Zolgensma, Rethymic, • RMAT designation (2): Breyanzi, Rethymic, Stratagraft	
EU	• Conditional approval (6): Zalmoxis, Holoclar, Tecartus, Abecma, Zolgensma, Zynteglo • PRIME designation (9): Kymriah, Yescarta, Zynteglo, Breyanzi, Skysona, Abecma, Tecartus, Zolgensma, Zynteglo • Approval under exceptional circumstance (1): Glybera	(14) Kymriah, Luxturna, Zynteglo, Yescarta, Alofisel, Zalmoxis, Strimvelis, Holoclar, Glybera, Tecartus, Libmeldy, Abecma, Skysona, Breyanzi, Zolgensma, Zynteglo
Japan	• Conditional approval (5): Collategene, Stemirac, HeartSheet, JACC, Delytact • Priority review (1): JACE • SAKIGAKE designation (2): Stemirac, Delytact	(5) JACE, Temcell HS, Delytact, Cural, Nepic
Korea	• Conditional approval (7): CreaVax-RCC, KeraHeal, Holoderm, Kaloderm, ChondronTM, RMS OssronTM, Immuncell-LC	(3) Immuncell-LC; Neuronata-R; Cupistem
Canada	• NoC/c) (4): Prochymal, Yescarta, Abecma, Tecartus • Priority review (1): Prochymal	(7) Kymriah, Yescarta, Prochymal, Abecma, Luxturna, Zolgensma, Tecartus
India	• Conditional approval (4): RMS Ossron, Chondron ACI, Apceden, Stempeucel	(1) Stempeucel in the EU

EU: European Union; ODD: Orphan Drug Designation; USA: United States.

[a]: No RMs have benefitted from accelerated approval pathways in Australia, New Zealand, Singapore, and China.

[b]: For countries (USA, EU, and Japan) with specific legislation for orphan drugs, the status of orphan drug designation was extracted from the regulatory authority website. For other countries (Korea, Canada, and India), the status of orphan drug designation was searched from the FDA, EMA, and PMDA websites.

1.2.8 POST-MARKET REQUIREMENTS FOR RMs APPROVED

In all countries and the EU, MA holders of conditional approval are required to collect more comprehensive data for the confirmation of their products' efficacy and safety after MA. Additional post-market requirements for products approved under accelerated programs are included in the United States, EU, Japan, Canada, and Australia. Stricter requirements may apply to RMs exclusively in some countries. In the United States, the CBER determines what types of post-market requirements (such as confirmatory clinical trials, patient registries, electronic health records, or other data collections) are necessary to confirm the benefits and safety of RMs approved under accelerated programs. In the EU, apart from general post-market requirements, post-authorization safety and efficacy (S&E) follow-up studies need to be put in place, which could comprise extension phases of pre-authorization trials, additional clinical trials, and/or other observational studies. In Japan, RMs receiving time-limited conditional MA must submit a standard MA application with post-market data collected from all patients treated during the pre-determined period of conditional approval. In Australia, some additional bio-vigilance responsibilities were required for higher-risk biological products, including a Risk Management Plan (RMP) and a Periodic Safety Update Report (PSUR). Health Canada has introduced enhanced post-market surveillance procedures for products having received NOC/c status, and it intends to use the systematic surveillance to enhance the current spontaneous adverse reaction reporting system.

1.2.9 RMs Sponsor Incentives

All nine investigated countries and the EU provide incentives for manufacturers to encourage investment in the development of innovative therapies. Non-financial incentives include scientific advice related to RM classification, manufacturing process, non-clinical study, and clinical trial design. Pre-submission consultation allows sponsors to obtain advice regarding the suitability and sufficiency of the MA application dossiers. Compliance with the advice recommendations given by regulators has a positive association, with a shorter MA review process, fewer major objections raised, and a higher possibility of positive MA decision-making (15). In order to mitigate any potential for financial conflict of interest, experts invited to advisory meetings for providing scientific advice and consultations are required to declare any conflict of interest related to the subject matter of the committee meeting. Other non-financial incentives include rolling review, allowing drug companies to submit completed sections of their MA applications; webinar/training programs for regulatory requirements education; and subject matter expert (SME) support/assistance. Financial incentives include fee reductions for scientific advice and MA activity, tax refunds or offsets for RM manufacturers, and a variety of governmental funding and/or grant programs (Table 1.14).

RMs receiving PRIME designation by the EMA, or RMAT designation by the FDA, or SAKIGAKE designation by the PMDA may benefit from more extensive interactions with regulators to obtain continuous guidance on development plans and regulatory strategies at the early development stage. Furthermore, RMs with PRIME designation, RMAT designation, or SAKIGAKE designation can expect to be eligible for expedited review and/or approval at the time of MA application. Apart from the regulatory support, RMs with SAKIGAKE designation have the potential to benefit from a 10% to 20% price premium, and RMs with PRIME designation can interface with additional stakeholders, such as health technology assessment (HTA) bodies.

TABLE 1.14

Expedited Programs and Incentives for Manufacturers

Incentives to encourage engagements

EU	• Non-financial: Scientific advice; protocol assistance; certification of quality and non-clinical data for SMEs • Financial: Fee reduction for request for scientific advice and certification procedure • Funding: Large-scale research initiative – RESTORE under the Horizon 2020
USA	• Non-financial: Advisory committee discussions; product-specific confidential inquiries during pre-submission process provided by OTAT; website/program for the developers of regenerative medicine called OTAT Learn; small business assistance program; Tissue Reference Group rapid inquiry programs (TRIP) • Financial: User fee waivers, reductions, and refunds for eligible biologicals • Funding: Common fund programs for regenerative medicine (RMP) initiated by the National Institutes of Health (NIH)
Japan	• Non-financial: PMDA consultation on R&D strategy and clinical trial; priority assessment consultation • Financial: Products designated as orphan RM are entitled to financial aid and tax relief; fee reduction for SMEs • Funding: World-leading innovative R&D on science and technology (FIRST)
Australia	• Non-financial: TGA advice on biologicals classification; SME assistance • Financial: Tax incentives (38.5–43.5% tax refund or offset) • Funding: Government initiatives targeted at stem cell research to facilitate research and development, such as the Biomedical Translational Fund, Medical Research Future Fund, and Austrade
Canada	• Non-financial: SME assistance • Funding: Federal programs, such as a grant from the California Institute for Regenerative Medicine (CIRM); networks aiming at regenerative medicine research, such as the Stem Cell Network (SCN) and Canadian Stem Cell Foundation
New Zealand	• Funding: Biomedical research funding jointly established by the National Natural Science Foundation of China (NSFC)
South Korea	• Non-financial: Pre-review system (prior screening); scientific advice on the pre-review of IND submission; biannual training program for educating on the regulatory requirements • Funding: Government budget assigned by Korea National Institute of Health (KNIH)
China	• Non-financial: Biotechnology is named as a strategic emerging industry under the 13th 5-year plan to prioritize its development, such as the creation of a platform for biotechnology innovation • Financial: Increase in investment for innovative biotechnology and reduced tax for biotechnology company • Funding: Biomedical research funding jointly established with Health Research Council (HRC) of New Zealand

TABLE 1.14 *(Continued)*
Expedited Programs and Incentives for Manufacturers

	Incentives to encourage engagements
Singapore	• Non-financial: Scientific and regulatory advice provided by Innovation Office; SME Health plus support
	• Funding: A*STAR's Biomedical Research Council funding for stem cell research; cell processing laboratory under cellular therapy program
India	• Non-financial: Focus support provided by DBT, such as the creation of innovative disease models and animal models
	• Funding: Biomedical innovation through policy initiatives, such as the National Biotechnology Development Strategy 2015–2020; funding provided by DBT

DBT: Department of Biotechnology; EU: European Union; IND: Investigational New Drug; PMDA: Pharmaceuticals and Medical Devices Agency; R&D: Research and Development; RMs: Regenerative Medicine; RMAT: Regenerative Medicine Advanced Therapy; OTAT: Office of Tissues and Advanced Therapies; RMAT: Regenerative Medicine Advanced Therapy; SME: Small and Medium-sized Enterprise; TGA: Therapeutic Goods Administration; USA: United States.

1.3 CHALLENGES AND RECOMMENDATIONS

1.3.1 DISCREPANCIES BETWEEN RM REGULATORY REQUIREMENTS

Inconsistency in regulatory requirements across countries remains a major hurdle for manufacturers to navigate the MA process harmoniously (16). Differences shown in the classification of RMs across the nine investigated countries and the EU revealed that tisagenlecleucel (Kymriah) was approved as a cellular therapy in Japan and Australia, yet it was approved as a gene therapy in the EU and United States. Such inconsistences across countries make it difficult for sponsors to decide which regulations should be applicable to their products. Global diversification in RM research regulations has been presented in aspects of research methods, ethical standards, and approval procedures, raising substantial concerns and posing barriers in the initiation of multicenter clinical trials (17). Such discrepancies are also reflected in the conflicting requirements for MA, not only in evidence generation methodologies but also in the divergent standards for efficacy and safety evaluations (14, 18). For example, the acceptance of the use of evidence generated from observational studies or patient registries varied across countries. The RMs granted conditional MA based on preliminary clinical data in one country may not be considered admissible in another country (19). Furthermore, eligibility criteria and application procedures for expedited approval programs differed internationally, which is illustrated by the absence of products receiving all three designations of RMAT designation, PRIME designation, and SAKIGAKE designation (20). This situation seems to be more prominent for academic institutions and start-up companies, who play a critical role in the current RM market, yet have limited experience and scarce financial resources to address complex regulatory requirements (21). This may lead to a delay in patient access because manufacturers are motivated to target large markets that have more predictable regulatory environments, leaving most of the less-developed world deprived from this new promising class of therapies.

1.3.2   THE SIGNIFICANCE OF POST-MARKET EVIDENCE GENERATION

1.3.2.1   Rising Concerns Around Expedited Approval Programs

Although regulators hold varying attitudes towards the balance between regulation stringency and flexibility, there is an overall tendency that regulators have more readily adopted conditional and adaptive approaches to stimulate the market approval of RMs (22). Therefore, small-sized, non-comparative studies using surrogate endpoints as primary outcomes may form the pivotal evidence for drugs under expedited approval pathways (23, 24). However, there is no proof that drugs approved under expedited programs provide significant advantages in efficacy or novelty compared with drugs approved through standard pathways (25). Drugs approved under expedited programs may fail to demonstrate clinical benefits in the confirmatory studies, thus experiencing MA withdrawal (26).

Accordingly, robust post-market evidence collection constitutes the key to the successful implementation of accelerated approval programs, to verify clinical benefits of RMs, and to mitigate the potential safety risk for patients. However, according to a review of the post-market surveillance studies for current approved ATMPs in the EU, post-market interventional studies generally adopted a methodology resembling the pre-market trials, with small patient numbers, short-term follow-up periods, and primary outcomes to answer the hypothesis from pre-marketing scenarios, rather than testing the real-world scenarios in a broader population (27). For cancer drugs receiving accelerated approval, improvement in surrogate endpoints, rather than overall survival, was reported in confirmatory studies, making the uncertainties in actual drug effects on patient-relevant endpoints unaddressed (28, 29).

1.3.2.2   Challenges in Post-Market Evidence Collection

Furthermore, there are growing concerns surrounding the fulfillment of post-market obligations. It was indicated that most post-market studies attached to conditional approvals were completed with a substantial delay and showed methodological discrepancies over time (13, 30). Information relating to randomization methods, comparator types, outcomes, and patient numbers were not sufficiently reported (31). From the perspective of conditional reimbursement, evidence submitted in the reassessment process was considered insufficient to justify the efficacy and safety (32), leading to the discontinuation of coverage with evidence development programs in some countries (33). This may be explained by multiple factors from the perspective of different stakeholders. For example, once a drug has reached the market, patients may express less willingness to be included in the post-market studies in the face of possibly being allocated to the control group (34). Manufacturers have little incentives to complete the post-market study in a timely manner. Post-market studies may even lead to a financial loss if new adverse effects are identified or products are restricted to a narrow subpopulation best responding to the treatment. Post-marketing requirements issued by regulators seem to be only briefly described without detailed information to understand the study design and the purposes (30).

Considering that RMs are generally associated with high uncertainty in long-term efficacy and safety, regulators have released a series of guidelines to increase the transparency and provide more instructions on carrying out post-market surveillance of RMs. The EMA has revised its "Guidelines on safety, efficacy, follow-up, and risk management of ATMPs," in which risk minimization measures as well as methodologies to design post-authorization follow-up studies are defined (35) (Table 1.15). The EC may impose financial penalties for the infringement of

TABLE 1.15

Post-Launch Studies for Approved RMs

Brand Name	Study Identifier	Study Design	Follow-up Duration	Sample Size	Status (as of Oct. 2020)
Provenge	NCT01727154	Observational, prospective case-only	52 weeks	139	Terminated
	NCT01306890	Registry study, observational, prospective cohort (PROCEED)	3 years	1976	Completed
Glybera	NCT03293810	Registry study, observational, prospective	15 years	16	Active, not recruiting
Imlygic	NCT02173171	Registry study, observational, prospective cohort	7 years	340	Enrolling by invitation
	NCT02910557	Post-marketing, prospective cohort study among patients treated in daily routine clinical practice	5 years	920	Recruiting
Strimvelis	NCT03478670	Long-Term, Prospective, Non-Interventional Follow-up of Safety and Efficacy, patient registry study (200195)	15 years	50	Enrolling by invitation

(Continued)

TABLE 1.15 *(Continued)*
Post-Launch Studies for Approved RMs

Brand Name	Study Identifier	Study Design	Follow-up Duration	Sample Size	Status (as of Oct. 2020)
Kymriah	NCT02445222	CAR-T Long Term Follow Up (LTFU) Study (PAVO)	15 years	1250	Recruiting
Luxturna	NCT03602820	Observational, multisite, non-randomized, prospective cohort, long-term safety and efficacy follow-up study (LTFU-01)	15 years	41	Active, not recruiting
	NCT03597399	Registry study, observational cohort	5 years	40	Active, not recruiting
Zynteglo	NCT02633943	Prospective, observational case-only, long-term follow-up study (LTF-303)	15 years	94	Enrolling by invitation
Zolgensma	NCT03421977	Observational, prospective Long-Term Follow-up Study for Patients From AVXS-101-CL-101 (START)	15 years	13	Active, not recruiting
Yescarta	NCT05041309	Long-term follow-up study for participants of kite-sponsored interventional studies treated with gene-modified cells	5 years	700	Enrolling by invitation
Carvykti	NCT05201781	Adult patients with relapsed or refractory multiple myeloma after four or more prior lines of therapy	15 years	228	Not recruiting yet

certain obligations specified in conditional approvals. Likewise, the FDA is developing a guideline titled "Long Term Follow-up After Administration of Human Gene Therapy Products," which stipulates that long-term follow-up (LTFU) observational studies must be in place for gene therapies presenting potential long-term risks to patients (36).

1.3.3 PROSPECTS AND RECOMMENDATIONS

The implementation of RM regulations, along with the introduction of accelerated regulatory pathways and attractive incentives, have become one of the most important catalysts for pharmaceutical companies to invest more in this field. The FDA has predicted that cell and gene therapies will be approved at a rate of 10 to 20 annually by 2025 (37). The United States, EU, Japan, and Korea will still be pioneers in this field with a wealth of RM research institutions (38), world-class manufacturing technologies, and a relatively well-developed RM regulatory system. Canada and Australia have made more efforts to clarify the regulatory requirements with several policy documents for the MA of RMs released recently, and both countries have expressed their determination to create a successful, internationally recognized RM industry in their future strategy and action plans (39, 40). New Zealand is developing a new regulatory scheme with a separate sector for cell products and tissue products in the draft "Therapeutic Products Bill" in order to improve the regulatory efficiency and accelerate the launch of RMs (41). China and India are also working on standardizing and reinforcing RM regulations in order to improve global competitiveness and address the use of unauthorized RMs in their countries. Overall, regulators worldwide regard RMs as a field for prioritization, which holds both great challenges and enormous potential for public health.

1.3.3.1 International Harmonization for RM Regulations

The discrepancies in regulatory requirements call for international coordination in order to standardize the terminology and to establish a universal regulatory pathway for RMs to align evidence requirements on the highest standard and allow international development to accelerate global access to innovative therapy. There already exists a number of international harmonization initiatives, such as the Gene and Cell Therapy Working Group of the International Pharmaceutical Regulators Forum (IPRF) (42). As claimed, the primary goals of the working group are to promote information sharing of best practices for the regulation of cell and gene therapies and reinforce the regulatory convergence. The International Council for Harmonization of Technical Requirements for Pharmaceuticals for Human Use (ICH) is also working on streamlining RM development and approval internationally, having established the Gene Therapy Discussion Group (GTDC) (43). The International Alliance for Biological Standards (IABS) also set the tone in the field of cell and gene therapy products as their priorities by active involvement in related initiatives in other organizations such as the WHO, ICH, and ISCT (44). As forerunners in the RM field, the EMA and the FDA are expected to work closely to exchange regulatory information related to RM guidelines and MA activities. Although a uniform platform could be challenging to establish, considering the diverse pharmaceutical regulation systems and public health needs in different countries, its advantage in terms of minimizing duplicative efforts, facilitating international integration, and increasing patient access without resorting to medical tourism must be highlighted. In December 2021, the WHO published "WHO Considerations on Regulatory Convergence of Cell and Gene Therapy Products (CGTPs)" (45) to promote convergency and encourage member states to strengthen their regulatory frameworks for cell and gene therapies. The priorities of the documents are outlined as follows:

- Clearly describe what the CGTPs are, describe how the subsets of HCTs and ATMPs are defined from this larger class, and provide definitions of key terminology relevant in this area. Examples will be provided in a subsequent document, as appropriate.
- Summarize the existing state of ATMPs that are approved or under development, including examples of challenges in the development and where solutions have been identified.
- Provide the key elements of a regulatory framework that supports the safety and effectiveness of CGTPs, including suggested regulatory controls for different risk categories of products covering key elements for adequate oversight spanning the entire product lifecycle from the investigational phase through post-market surveillance.
- Develop a proposal for how the regulatory framework for the risk categories could be implemented in countries with different levels of regulatory maturity.
- Provide an annotated bibliography to highlight key references relevant to the manufacture, product development, and regulation of ATMPs.

The WHO goal is to promote regulatory convergence for CGTPs to facilitate development and access to these novel products for patients in all regions of the world. In addition, the aim is to increase the safety of patients treated with CGTPs by preventing exploitation of those jurisdictions with inadequate regulations in place for the safe oversight of such novel products.

1.3.3.2 More Clarification and Transparency in the Regulatory Activities

Seeking scientific advice from regulators will be beneficial for manufacturers to gain valuable feedback prior to MA submission, yet more transparencies are necessary to prevent the regulators' ultimate decisions from being influenced by such early interactions. Provision of consultation with exchange of regulatory fees carries an inherent risk in obtaining MA (46), possibly due to the fear of damage to an authority's reputation if the given advice is rejected in the MA evaluation process. It is important for regulators to separate between the experts responsible for providing scientific advice and those involved in the subsequent MA evaluation of the same products. The advice given in the pre-submission meeting and the experts involved in the pre-submission activities should be accessible to the public, with the exception of involvement in sensitive business plans (47). Another proposal argues that the MA decision may be more impartial if regulatory agencies' incomes are independent of the fees from industry and are reliant on the national budgets (48).

1.3.3.3 Strategies to Ensure That Post-Market Scientific Obligations Are Fulfilled

For the successful implementation of expedited approval programs to promote patient access to innovative drugs satisfying unmet needs, the criteria of these programs should introduce the demonstration of a new drug's therapeutic advances compared with the best available treatment options (46). Post-market requirements should highlight the validation of surrogate endpoints in predicting the actual clinical benefits (29). RWE could be considered when the confirmatory RCT is unfeasible to conduct, while strategies to mitigate the potential selection bias and confounding factors must be put in place (49). Regulators should closely monitor the post-market activities and apply penalties in the case of non-compliance with the scientific obligations. All these issues related to post-market studies would be of more significance for RMs, considering that its curative effects claims could only be examined in the post-market studies with long enough study durations. When MA holders are resistant to properly developing and abiding by the post-approval commitment, an independent third body should be engaged to perform the due study at the expenses of the MA holders, with no chance for the MA holders to interfere with the study. MA holders should be requested to sign this commitment before MA is issued. It is not possible to speculate on public health, especially when a drug is launched and reimbursed, as it is difficult to reverse the decision without robust evidence (50). For example, among the total 168 cancer drugs receiving accelerated approval, only 19 of them experienced approval withdrawal by FDA (Table 1.16).

TABLE 1.16
Withdrawal of Cancer Accelerated Approvals

Drug Name	Accelerated Approval (AA) Indication	Accelerated Approval Date	Withdrawal Date
Keytruda (pembrolizumab)	Metastatic SCLC with disease progression on or after platinum-based chemotherapy and at least one other prior line of therapy	6/17/2019	3/30/2021
Tecentriq (atezolizumab)	In combination with paclitaxel protein-bound for unresectable locally advanced or metastatic triple-negative breast cancer whose tumors express PD-L1 (PD-L1 stained tumor-infiltrating immune cells of any intensity covering = 1% of the tumor area), as determined by an FDA-approved test	3/8/2019	10/6/2021

(Continued)

TABLE 1.16 *(Continued)*
Withdrawal of Cancer Accelerated Approvals

Drug Name	Accelerated Approval (AA) Indication	Accelerated Approval Date	Withdrawal Date
Copiktra (duvelisib)	Treatment of adult patients with relapsed or refractory follicular lymphoma (FL) after at least two prior systemic therapies	9/24/2018	12/17/2021
Opdivo (nivolumab)	Metastatic SCLC with progression after platinum-based chemotherapy and at least one other line of therapy	8/16/2018	12/29/2020
Keytruda (pembrolizumab)	For patients with recurrent or locally advanced or metastatic gastric or GEJ adenocarcinoma whose tumors express PD-L1 [CPS ≥1] as determined by an FDA-approved test, with disease progression on/after two or more prior lines of therapy, including fluoropyrimidine- and platinum-containing chemotherapy, and if appropriate, HER2/NEU-targeted therapy	9/22/2017	2/4/2022
Opdivo (nivolumab)	Hepatocellular carcinoma previously treated with sorafenib	9/22/2017	7/23/2021
Imfinzi (durvalumab)	Locally advanced or metastatic urothelial carcinoma that progressed during or following platinum-containing chemotherapy or within 12 months of neoadjuvant or adjuvant treatment with platinum-containing chemotherapy	5/1/2017	2/19/2021
Lartruvo (olaratumab)	In combination with doxorubicin for adults with soft tissue sarcoma with a histologic subtype for which an anthracycline-containing regimen is appropriate and which is not amenable to curative treatment with radiotherapy or surgery	10/19/2016	2/25/2020
Tecentriq (atezolizumab)	Locally advanced or metastatic urothelial carcinoma that progressed during or following platinum-containing chemotherapy or within 12 months of neoadjuvant or adjuvant treatment with platinum-containing chemotherapy	5/18/2016	4/13/2021
Farydak (panobinostat)	In combination with bortezomib (BTZ) and dexamethasone (DEX) for the treatment of patients with multiple myeloma (MM) who have received at least two prior regimens, including BTZ and an immunomodulatory agent	2/23/2015	3/24/2022

TABLE 1.16 *(Continued)*
Withdrawal of Cancer Accelerated Approvals

Drug Name	Accelerated Approval (AA) Indication	Accelerated Approval Date	Withdrawal Date
Zydelig (idelalisib)	For the treatment of relapsed follicular B-cell non-Hodgkin lymphoma (FL) in patients who have received at least two prior systemic therapies and relapsed small lymphocytic lymphoma (SLL) in patients who have received at least two prior systemic therapies	7/23/2014	2/18/2022
Marqibo (vincristine sulfate liposomal)	Adults with Philadelphia (PH) chromosome–negative (–) ALL in second relapse or greater relapse or whose disease has progressed following two or more treatment lines of antileukemia therapies	8/9/2012	5/2/2022
Istodax (romidepsin)	Peripheral T-cell lymphoma in patients who have received at least one prior therapy	6/16/2011	7/30/2021
Oforta (fludarabine phosphate)	For adults with B-cell CLL that has not responded to or progressed during or after treatment with at least one standard alkylating agent–containing regimen	12/18/2008	12/31/2011
Avastin (bevacizumab)	In combination with paclitaxel for patients who have not received chemotherapy for metastatic HER2-negative breast cancer	2/22/2008	11/18/2011
Bexxar (tositumomab and iodine i131 tositumomab)	For patients with relapsed or refractory low-grade follicular or transformed CD20+ NHL who have not received rituximab	12/22/2004	10/23/2013
Iressa (gefitinib)	As monotherapy for locally advanced or metastatic NSCLC after failure of platinum-based and docetaxel chemotherapy	5/5/2003	4/25/2012
Mylotarg (gemtuzumab ozogamicin)	For patients with CD33+ AML in first relapse 60 years of age or older and who are not candidates for cytotoxic chemotherapy	5/17/2000	11/28/2011
Celebrex (celecoxib)	To reduce the number of adenomatous colorectal polyps in familial adenomatous polyposis patients, as an adjunct to usual care	12/23/1999	6/8/2012

Note: Data extracted from FDA website up until 03/05/2022.

1.3.3.4   Patient Registry Platforms for Rare Diseases

In post-marketing phases, a rare disease registry is powerful to collect the comprehensive RWE of new treatments. Compared to clinical trials, it can examine the effectiveness and safety of products in routine practice and expand to wider and diverse populations. Therefore, the uncertainties around the generalizability of clinical evidence in initial MA to a real-world setting could be better addressed (51, 52). Additionally, a centralized treatment registry platform would be an efficient approach to facilitate post-market evidence collection while reducing the recruitment burdens within a small patient group and avoiding duplicative efforts from different developers (53).

However, there are significant challenges in the use of patient registries for post-market evidence collection, such as the lack of standard methodologic protocols to design, report, and analyze data gathered from patient registries and the limited use of patient registries outside of the original countries establishing them. In an acknowledgement of the existing challenges (54), the EMA launched initiatives for patient registries in 2015, with the intention to create a European Union–wide framework on patient registries. This initiative aims to facilitate the collaborations between all relevant stakeholders related to rare diseases, consisting of patients, physicians, academic institutions, pharmaceutical companies, and national competent authorities (55).

One successful case is the European Society for Blood and Marrow Transplantation (EBMT) cell therapy registry. It has received a regulatory qualification from the EMA as a suitable platform for pharmacoepidemiologic studies for regulatory purposes, concerning CAR-T cell therapies in March 2019 (56). This ensures all the post-approval commercial administration of Kymriah and Yescarta will be reported to the EBMT in the EU, thus substantially streamlining the data collection on the long-term safety and efficacy of drugs. Another example is SMArtCARE, which is a prospective, multicenter, non-randomized registration and outcome study. SMArtCARE (57) collects longitudinal data on all available SMA patients independent of their actual treatment regimen as a disease-specific SMA registry. SMArtCARE could serve as an online platform for SMA patients seen by healthcare providers in Germany, Austria, and Switzerland. The evaluation of real-life outcome data in a broad spectrum of SMA patients will lead to a better understanding of the natural history and examine the influence of drug treatment, including Zolgensma, Spinraza, and Evrysdi. This will largely streamline the collection of RWE on the effectiveness and safety of Zolgensma, as well as the comparative evidence between all available treatments, leading to a more robust assessment and better-informed decision-making.

1.4   CONCLUSION

The uniqueness of RMs and their potential to prevent the onset of disease and delay disease progression and even cure some potentially fatal conditions renders a customized approach for their evaluation and regulation critical and necessary.

Although specific RM legislation and frameworks have only been made available in the EU, United States, Japan, and Australia, regulators in most countries have given additional considerations in evidence assessment by considering the specificities of RMs. Moreover, all investigated countries and the EU recognized the importance of establishing expedited pathways to accelerate the approval of RMs. International coordination to reach an agreement clarifying the minimal regulatory requirements for the MA of RMs is critical to reduce the exhaustive administration burden for manufacturers and to ensure faster patient access to innovative therapies. Expedited approval programs are not risk-free, especially considering there still exists a great deal of uncertainty on the sustainable benefit and the future safety at the time of MA. Post-approval studies are crucial to confirm the actual efficacy and safety of innovative RMs. Deviation and delays to post-approval commitment should not be tolerated, as it is considered as a public health priority. Regulation should allow an independent third party to be engaged to perform the requirement at the expense of the manufacturer when needed. RMs represent an immense hope for suffering patients and their caregivers, but should also be handled as novel pharmaceuticals associated with high uncertainty and scarce experience thus far.

2 Health Technology Assessment of RMs

RMs have offered new treatment options for disease areas with limited treatment availability or lack of effective treatments rather than simply controlling or relieving symptoms. The novel mechanism of action of RMs has contributed to the unique advantages allowing for the underlying causes of the diseases to be corrected, thus, potentially relieving patients from lifelong treatments through its one-time administration.

For example, two chimeric antigen receptor (CAR) T-cell therapies, tisagenlecleucel (Kymriah) and axicabtagene ciloleucel (Yescarta) received market authorization (MA) from both the EU and the United States, considering the innovations' abilities to reprogram patients' own cells to attack cancer, with more precision possible due to the specific protein recognition of and binding to the tumor cells. Voretigene neparvovec (Luxturna), by introducing a functional gene copy to compensate for gene mutations, was associated with important clinical benefits compared with the standard of care in improving the patients' vision and quality of life.

It is worth noting that single-arm studies with limited patient numbers, short follow-up durations, and reporting only surrogate endpoints, instead of well-designed and robust randomized controlled trials (RCTs), constituted the pivotal evidence for several approved cell and gene therapies. This reflects the regulators' acceptance of a degree of uncertainty in the cases of such breakthrough innovations with positive benefit-harm ratios.

In contrast to the regulator's enthusiasm to facilitate timely market approval based on limited evidence, health technology assessment (HTA) bodies hold more conservative attitudes towards ATMPs. HTA bodies concluded that the limited clinical evidence increased the uncertainties surrounding the curative potential, the magnitude and the durability of clinical benefits, and the potential unfavorable effects in the long run, which raised barriers for payers to reimburse ATMPs within tight budgets (58).

The fundamental gap between payers and regulators is related to their different mandates. Regulators are responsible for safeguarding the public. As such they consider the quality and the direction of the benefit-risk ratio. If the benefit-risk ratio is positive, they will endorse the drug for MA. Regulators are under strong pressure from the parliament to accelerate patients' access to new innovative therapies. Payers, on the other side, must consider the magnitude of the benefit over the magnitude of the risk versus currently used interventions. The magnitude will inform payers' willingness to pay for a new intervention. With immature data, payers may experience massive uncertainty to appreciate the magnitude of benefit

DOI: 10.1201/9781003366676-2

and risk of a new drug. Payers are accountable for the budget set by the parliament, which often is disconnected from the parliament's public health objective. They manage governmental taxpayer budgets, employers' budgets, nonprofit or for-profit private budgets, and eventually patients' out-of-pocket payments. The perspective is totally different, and the strategy and decisions are misaligned.

The objective of this chapter is to investigate the HTA decisions of approved RMs made and further analysis of the limitations of the clinical evidence for RMs that are criticized by the HTA bodies studied.

2.1   COUNTRY-SPECIFIC HTA DECISIONS FOR INDIVIDUAL PRODUCTS

2.1.1   GLYBERA (ALIPOGENE TIPARVOVEC)

Glybera, used for the treatment of adult patients with familial lipoprotein lipase (LPL) deficiency who have severe or multiple pancreatic crises despite a low-fat diet, was reviewed by the HAS in France and by the Germany Federal Joint Committee (G-BA) (Table 2.1).

- **Limitations of the study design**

The pivotal evidence for the assessment of Glybera is two multicenter, single-arm, open-label studies: CT-AMT-011–01 and CT-AMT-011–02. Both the G-BA and HAS considered that pivotal evidence based on non-comparative studies was too poor to make a robust assessment in terms of product benefit.

The clinical relevance of the chosen primary endpoint, reduction in the triglyceride level, was questioned by both G-BA and HAS. Regarding the study population, HAS considered that a very small number of patients were included in the clinical trials in relation to the potential patient population eligible for Glybera treatment, while G-BA determined that only a third of the patients enrolled in the clinical trials corresponded to the indication approved. Furthermore, HAS pointed out that the absence of information on the dietary regimen increased the uncertainty that the reduction in the triglyceride level was caused by Glybera treatment or simply by improved compliance with a better dietary regimen (Table 2.1).

TABLE 2.1
Evidence Limitations of Glybera

	Study design	Study Population	Comparator	Outcome Assessment	Analysis Method	Transferability/ Generalizability	Other Confounding Factors	Indirect Comparison
France	√	√	NA	√	×	×	√	×
Germany	√	√	NA	√	×	×	×	×

2.1.2 KYMRIAH (TISAGENLECLEUCEL)

Kymriah, indicated for the treatment of B-cell acute lymphoblastic leukemia (ALL) and for the treatment of relapsed or refractory diffuse large B-cell lymphoma (DLBCL), has been evaluated in five countries (UK, France, Germany, Scotland, and United States). It has received positive recommendations for both indications in four countries (UK, France, Germany, and United States). In Scotland, Kymriah was accepted for the treatment of B-cell ALL, yet failed to be accepted for the treatment of DLBCL (Table 2.2).

- **Limitations of the study design**
  - **DLBCL indication**

The pivotal evidence for the assessment of Kymriah for DLBCL is the JULIET study (NCT02445248), which is an international, phase 2, single-arm study. All HTA agencies criticized that the lack of a comparative group made the precise estimation of the treatment benefit compared with salvage chemotherapy difficult to establish.

HAS underlined that the representativeness of study population was not guaranteed; accordingly, the transferability of study results to all patients with an advanced stage of disease who were relapsed or refractory to several lines of treatment was not ensured. Besides, the high drop-out rate seen in the study made it difficult to draw meaningful conclusions. HAS, G-BA, and Scottish Medicines Consortium (SMC) assessed that the use of an efficacy analysis set (EAS) ignored the impact of a waiting time before treatment and bridging treatment, potentially leading to the overestimation of the treatment benefit of Kymriah. The National Institute for Health and Care Excellence (NICE) considered that the analysis of overall survival based on Kaplan-Meier curves was highly uncertain due to few patient numbers from month 20 onwards. HAS, G-BA, Tandvårds- och läkemedelsförmånsverket (TLV), and SMC criticized that the impact of confounding factors was not fully investigated, including the waiting time until the availability of products, as well as the bridging treatment prior to Kymriah treatment.

Indirect comparisons were made using the data obtained from three major sources: 1) the SCHOLAR study pooling data from two phase 3 clinical trials and two observational cohort studies, 2) the CORAL extension trial compared two salvage chemotherapy regimens with and without rituximab, and 3) the Hematological Malignancy Research Network (HMRN). However, the capacities of indirect comparisons to draw meaningful conclusions were limited due to the differences in the baseline characteristics of study populations, outcome assessments, and previous treatment compared with the JULIET study. HAS stated that the SCHOLAR-1 study was not specified as a control group in the protocol of the JULIET study. Additionally, the SCHOLAR-1 study was subject to inherent limitations in terms of meta-analysis, such as the selection bias and confounding bias.

- **B-cell ALL indication**

TABLE 2.2
Evidence Limitations of Kymriah

	Study Design	Study Population	Comparator	Outcome Assessment	Analysis Method	Transferability/ Generalizability	Other Confounding Factors	Indirect Comparison
• Kymriah – DLBCL								
England	✓	×	NA	×	✓	×	×	✓
France	✓	✓	NA	×	✓	✓	✓	✓
Germany	✓	×	NA	×	✓	×	✓	✓
Sweden	✓	×	NA	×	×	×	×	✓
Canada	✓	×	NA	×	×	×	×	×
Scotland	✓	×	NA	×	✓	×	✓	✓
• Kymriah – B-cell ALL								
England	✓	×	NA	×	×	✓	×	✓
France	✓	✓	NA	×	✓	✓	×	✓
Germany	✓	×	NA	×	✓	×	✓	✓
Sweden	✓	×	NA	×	×	✓	✓	✓
Canada	✓	×	NA	×	×	×	×	×
Scotland	✓	×	NA	×	×	×	×	✓
USA	✓	×	NA	×	✓	×	×	✓

The pivotal evidence for the assessment of Kymriah for ALL is the ELIANA study (NCT02435849), which is an international, phase 2, single-arm study. All HTA agencies criticized that the lack of a comparative group made the precise estimation of the treatment benefit compared with salvage chemotherapy difficult to establish.

HAS was concerned that the study population was heavily pre-treated and very few patients with primary refractory or relapsed disease were included, which limited the transferability of study results to a real-world setting. NICE and TLV also questioned the data transferability in clinical practice considering that the efficacy of Kymriah after blinatumomab treatment was not investigated, and the time interval between leukapheresis and relapse showed in the studies was longer than expected in clinical practice. HAS, G-BA, and ICER pointed out the intention to treat (ITT) analysis was more appropriate than full analysis set (FAS) analysis. FAS analysis had limited capacities to capture the impact of other confounding factors and excluded the patients who did not receive the treatment due to manufacturing failures, death prior to infusion, and adverse effects. G-BA and TLV assessed the existence of confounding factors, especially the impact of bridging chemotherapies, which increased the uncertainty in the analysis of clinical benefits. Besides, G-BA criticized the absence of detailed information on patient flow after the administration of Kymriah as well as the median duration for the follow-up of adverse events.

Indirect comparison with other chemotherapies was made using the clinical evidence mainly from seven studies, while HAS, G-BA, and TLV concluded that the indirect comparison submitted was not relevant enough to ensure a valid comparison between Kymriah and other chemotherapies. The differences in the patient characteristics between studies were the most criticized, including the inclusion criteria for studies, disease status, number of previous treatments, the time since the last relapse, outcome reports, and the length of follow-up duration. Several important prognosis factors were not adequately adjusted; thus, there was a possibility that patients had a more progressive disease status in the indirect comparative group than in the Kymriah group, and this could not be excluded. HAS and G-BA considered that the study serving as a control group was subject to significant selection bias because a systematic analysis of the different biases for each study was not provided and the choice of included studies was not pre-specified in the study protocol. Additionally, indirect comparisons also suffered from considerable uncertainty given the limitations for included studies regarding the small patient number, short follow-up duration, and the incomplete information on the patient baseline characteristics (Table 2.2).

2.1.3 Luxturna (Voretigene Neparvovec)

Luxturna for the treatment of inherited retinal dystrophies caused by *RPE65* gene mutations has been evaluated in the United States, the England, France, Germany, and Scotland (Table 2.3).

TABLE 2.3
Evidence Limitations of Luxturna

	Study Design	Study Population	Comparator	Outcome Assessment	Analysis Method	Transferability/ Generalizability	Other Confounding Factors	Indirect Comparison
England	√	√	×	√	×	×	×	×
France	×	√	×	×	×	√	×	×
Germany	√	×	×	√	×	×	×	×
USA	×	×	×	√	×	×	×	×
Scoltand	√	√	×	√	×	×		

- **Limitations of the study design**

The pivotal evidence for the assessment of Luxturna is Study 301, which is a phase 3, multicenter, open-label, randomized control study comparing Luxturna with standard of care (SoC). NICE and HAS noticed that only patients diagnosed with Leber congenital amaurosis (less common but more progressive diseases) with biallelic *RPE65* mutations were included in the study, but it was considered acceptable to extrapolate the results from the study population to all inherited retinal dystrophies with biallelic *RPE65* mutations given the difficulties in the patient enrollment for ultra-rare diseases. SMC mentioned that available evidence is based on data from a small number of heterogeneous patients only with a range of RPE65 mutations and levels of disease progression and baseline MLMT scores. The overall treatment effect may not be generalisable in terms of benefit:risk ratio in individual patients. However, HAS noted that the transferability of study results to clinical practice was questionable due to the absence of long-term evidence. Likewise, clinical experts in NICE stated that the clinical relevance of study results to clinical practice was difficult to predict in patients with less severe types of diagnosis such as retinitis pigmentosa (RP).

The study endpoints were subjected to the most criticisms: ICER considered that there were uncertainties regarding how the improvement in the multiluminance mobility testing (MLMT) scored (primary endpoints) could influence the daily activities in a real-world setting; SMC is concerned that there might be ceiling effect affecting the ability of the MLMT to detect further change over time, and the actual treatment effect may therefore be underestimated. NICE acknowledged that secondary endpoints, visual field and visual acuity, were usually considered unreliable because of intertest variability and lacked the capacity to capture the disease symptoms such as night blindness; and G-BA considered that the full-field light sensitivity threshold (FST) test and perimetry remain significant uncertainties regarding operationalization and repeatability (Table 2.3).

2.1.4 Imlygic (Talimogene Laherparepvec)

The pivotal evidence for the assessment of Imlygic is Study 005/05, which is a phase 3, multicenter, open-label, randomized controlled study comparing Imlygic with granulocyte macrophage colony-stimulating factor (GM-CSF). Imlygic for the treatment of unresectable, regionally, or distantly metastatic melanoma has been evaluated in England and Germany. NICE considered it to be cost-effective when compared with available treatments other than immunotherapies, while G-BA determined Imlygic to have no added benefit due to inappropriate comparators (Table 2.4).

- **Limitations of the study design**

NICE expressed concerns about the potential for bias arising from the open-label study design. Besides, the differences in the withdrawal rates between two groups and the use of a non-validated primary endpoint (durable response rate) made it difficult to interpret the efficacy results. The use of GM-CSF as a comparator was subjected to intensive criticisms: NICE considered it clinically ineffective and was not used in the clinical practice in the National Health Service (NHS), while G-BA criticized the fact that GM-CSF did not concur with the appropriate comparative therapy (ACT) suggested, making the assessment of added benefits versus ACT impossible.

Indirect comparison with ipilimumab was conducted using the data from a systematic review and meta-analysis, pooling the relevant clinical trials. However, NICE determined that the influence of other prognostic factors might be underestimated, potentially resulting in an overestimate of the efficacy of Imlygic compared with ipilimumab (Table 2.4).

2.1.5 Strimvelis

Strimvelis for the treatment of severe combined immunodeficiency (SCID) due to adenosine deaminase deficiency (ADA) has been evaluated in England only (Table 2.5).

TABLE 2.4
Evidence Limitations of Imlygic

	Study Design	Study Population	Comparator	Outcome Assessment	Analysis Method	Transferability/ Generalizability	Other Confounding Factors	Indirect Comparison
England	√	×	√	√	×	×	√	√
Germany	√	×	√	×	×	×	×	×

TABLE 2.5
Evidence Limitations of Strimvelis

	Study Design	Study Population	Comparator	Outcome Assessment	Analysis Method	Transferability/ Generalizability	Other Confounding Factors	Indirect Comparison
England	√	×	√	×	×	×	×	√

• **Limitations of the study design**

The pivotal evidence for the assessment of Strimvelis is AD1115611, which is a phase 1/2, multicenter, single-arm, open-label, interventional study. The results for several clinical trials were reported together as an "integrated population," and the pivotal evidence was derived from the study AD1115611, one open-label, single-arm, sequential study. The limitations of the clinical evidence were mainly around the small patient number and insufficient evidence for the comparator. Indirect comparison was made with Hassan et al.'s study, the largest data source on outcomes for patients with ADA-related SCID having a hematopoietic stem cell transplantation (HSCT). However, the small patient numbers and differences in reported outcomes made the indirect comparison between Strimvelis and HSCT difficult (Table 2.5).

2.1.6 YESCARTA (AXICABTAGENE CILOLEUCEL)

Yescarta for the treatment of DLBCL and primary mediastinal large B-cell lymphoma (PMBCL) has been evaluated in five countries (England, Scotland, France, Germany, and the United States) (Table 2.6).

• **Limitations of the study design**

The pivotal evidence for the assessment of Yescarta is the ZUMA-1 study, which is a phase 1/2, multicenter, single-arm, open-label study. All HTA bodies considered that the lack of comparative data made the assessment of the exact benefit size compared with salvage chemotherapies difficult to establish.

HAS underlined that the representativeness of the study population was not guaranteed; accordingly, the transferability of study results to all patients with an advanced stage of disease who were relapsed or refractory to several lines of treatment was not ensured. SMC questioned that no subgroup analysis for only PMBCL patients but for the mixing population of PMBCL and DLBCL-transformed follicular lymphoma (TFL) was conducted; additionally, the impact of the bridging therapy on the treatment effects was unknown. SMC and TLV considered that assessment of the overall response rate (ORR) could be subject to

TABLE 2.6
Evidence Limitations of Yescarta

	Study Design	Study Population	Comparator	Outcome Assessment	Analysis Method	Transferability/ Generalizability	Other Confounding Factors	Indirect Comparison
France	√	×	NA	×	×	√	×	√
Germany	√	×	NA	×	×	×	×	√
England	√	×	NA	×	×	×	×	√
Scotland	√	√	NA	×	×	×	√	√
Sweden	√	×	NA	√	×	×	×	√
USA	√	×	NA	√	×	×	×	√

bias due to the open-label study design, while ICER questioned the relevance of ORR for long-term overall survival.

An indirect comparison made between the ZUMA-1 study and SCHOLAR study was submitted to all HTA agencies. It was considered appropriate for decision-making in NICE, HAS, SMC, and ICER, while G-BA and TLV concluded that a robust evaluation of the effect size was difficult due to the uncertainty of indirect comparison. Considerable heterogeneity in the study population between the ZUMA-1 and SCHOLAR studies existed, and the statement justifying the choice of prognostic factors was not available. Additionally, the SCHOLAR study was associated with inherent limitations: incomplete data on the patient characteristics due to the retrospective nature of study and the potential selection bias for the studies included in the meta-analysis.

2.1.7 ZYNTEGLO (BETIBEGLOGENE AUTOTEMCEL)

Zynteglo (betibeglogene autotemcel) for transfusion-dependent beta-thalassemia (TDT) was evaluated by HAS, G-BA, and NICE. However, in the NICE draft guidelines released in February 2021, Zynteglo was rejected because the cost estimate was considerably higher than what NICE normally considers an acceptable use of NHS resources, at £1,450,000 per patient. Thus Zynteglo was suspended in the UK, considering that Bluebird (the developer of Zynteglo) withdrew its marketing application from the Medicines and Healthcare Products Regulatory Agency (MHRA) and does not intend to commercialize the therapy in the UK. In France, Zynteglo was only recommended in TDT patients aged between 12 and 35, and was not recommended for patients older than 35 years old, considering that clinical evidence for patients this age was lacking. G-BA evaluated the added benefit of Zynteglo as "non-quantifiable" due to the limited clinical evidence preventing any firm conclusions from being drawn.

TABLE 2.7

Evidence Limitations of Zynteglo

	Study Design	Study Population	Comparator	Outcome Assessment	Analysis Method	Transferability/ Generalizability	Other Confounding Factors	Indirect Comparison
France	√	×	×	×	×	×	×	×
Germany	√	√	×	√	×	×	×	×

- **Limitations of the study design**

The pivotal evidence for the assessment of Zynteglo is three single-arm, open-label studies: HGB-204, HGB-205, and HGB-207. The uncertainties relative to the maintenance of efficacy in the medium and long term were criticized by HAS and G-BA. Moreover, HAS stressed that there are uncertainties relative to safety, including the potential risk of immunization, insertional mutagenesis, and the impact on fertility, which must be closely monitored in the long-term follow-up studies. G-BA considered that owing to the small number and limited selection of patients in the studies due to the study's comprehensive inclusion and exclusion criteria, there are still uncertainties regarding the proportion of patients in the therapeutic indication who achieve transfusion independence after treatment with Zynteglo. For the outcome assessment, G-BA concluded that it remains unclear to what extent the transfusion independence exhibited in most patients prevents the complications that may arise following previous routine transfusions with red cell concentrates and to what extent iron chelation therapy is still necessary in patients who have achieved transfusion independence (Table 2.7).

2.1.8   LIBMELDY (ATIDARSAGENE AUTOTEMCEL)

Libmeldy (atidarsagene autotemcel) for children with metachromatic leukodystrophy (MLD) was assessed in France and Germany. The HTA bodies delivered sperate evaluations for the two subpopulations targeted by Libmeldy.

- For the subgroup of MLD symptomatic children with early clinical manifestations of the disease who still can walk independently and before the onset of cognitive decline with the early juvenile form (onset from 30 months to 6 years of age inclusive), the France HAS determined that Libmeldy had insufficient *Le service médical rendu* (medical benefit, SMR), and hence was not recommended for reimbursement. Germany G-BA assessed that the added benefits of Libmeldy are non-quantifiable due to the substantial limitations in the clinical evidence.
- For the subgroup of MLD asymptomatic children without clinical manifestation of the disease, whether in terms of motor, cognitive, and/or behavioral impairment, with the late infantile form (manifesting itself before 30 months) or early juvenile form (manifesting itself between

30 months and 6 years inclusive), HAS determined that Libmeldy had moderate *l'amélioration du service médical rendu* (added medical benefits, ASMR III), and G-BA determined that Libmeldy is associated with a hint of a considerable additional benefit.

- **Limitations of the study design**

The pivotal evidence for the assessment of Libmeldy is Study 201222, which was a non-randomized, open-label study. The uncertainties relative to maintenance of efficacy in the medium and long term was criticized by HAS and G-BA. Moreover, HAS stressed that there are uncertainties relative to safety, including the potential risk of immunization, insertional mutagenesis, and the impact on fertility, which must be closely monitored in the long-term follow-up studies (Table 2.8).

2.1.9 ZOLGENSMA (ONASEMNOGENE ABEPARVOVEC)

Zolgensma (onasemnogene abeparvovec) for spinal muscular atrophy (SMA) was assessed in England, Scotland, France, and the United States. In all countries, Zolgensma received a positive opinion for the indication of type I SMA and type II SMA with up to three copies of *SMA2* genes. It was recommended under a managed market entry agreement in England and Scotland. In the re-evaluation report, the France HAS concluded a "moderate" amount of added clinical benefits only in type I SMA and type II SMA with one to two copies of *SMA2* genes, which is more restricted compared to the initial assessment.

- **Limitations of the study design**

The pivotal evidence for the assessment of Zolgensma is the STR1VE study, which is a phase 3, multicenter, single-arm, open-label study. All HTA bodies determined that there are substantial uncertainties in the long-term clinical effectiveness and safety due to the short follow-up duration and small patient number. To compensate for the limitation of the single-arm trial, comparisons with natural history studies were made. England NICE considered that the natural history studies all had limitations, including a high proportion of people who have a tracheostomy, unlike best supportive care in the NHS. This was because the studies

---

TABLE 2.8

Evidence Limitations of Libmeldy

	Study Design	Study Population	Comparator	Outcome Assessment	Analysis Method	Transferability/ Generalizability	Other Confounding Factors	Indirect Comparison
France	√	×	NA	×	×	×	×	×
Germany	√	×	NA	×	×	×	×	×

TABLE 2.9

Evidence Limitations of Zolgensma

	Study Design	Study Population	Comparator	Outcome Assessment	Analysis Method	Transferability/ Generalizability	Other Confounding Factors	Indirect Comparison
England	√	×	NA	×	×	√	×	√
France	√	×	NA	×	×	×	×	√
US	√	×	NA	×	×	√	√	×
Scotland	√	×	NA	×	×	×	×	√

were set in the United States, where tracheostomy is more commonly used in this population, undermining the generalizability of study results in the England setting. Scotland SMC also expressed concerns towards indirect comparisons, for example, differences in baseline characteristics that were not able to be matched in the matching adjusted indirect comparison (MAIC) and not matched in the naïve indirect treatment comparison (ITC), some clinically relevant outcomes were not included, data immaturity, and for the results, confidence intervals (CIs) were generally wide and spread out. The US ICER considered that the narrow eligibility criteria of trials and the limited sample size raise concerns about the generalizability of results to the wider population of patients with SMA. The ineligible or otherwise unselected patients are likely more severely ill, experience different or additional comorbidities (e.g., scoliosis), or have a different genetic profile than those selected for the clinical trials (Table 2.9).

2.1.10 TECARTUS (BREXUCABTAGENE AUTOLEUCEL)

Tecartus was assessed in England, France, Germany, and Scotland. All four countries recommended reimbursement, and England and Scotland required that managed entry agreements must be followed.

- **Limitations of the study design**

The pivotal evidence for the assessment of Tecartus is the ZUMA-2 study, which is a phase 2, multicenter, single-arm, open-label study. All HTA bodies determined that there are substantial uncertainties in the long-term clinical effectiveness and safety due to the short follow-up duration and small patient number. The limitations in the indirect comparison were noticed in Germany and Scotland. G-BA determined that either the indirect comparison using MAIC based on a meta-analysis with eight external control studies or the indirect comparison with SCHOLAR-2 is appropriate due to the discrepancy in the study characteristic (i.e., inclusion criteria, patient flow, baseline characteristics, and operationalization of endpoints) and missing information on the study details. SMC critiqued the indirect comparison report for its lack of safety and

TABLE 2.10
Evidence Limitations of Tecartus

	Study Design	Study Population	Comparator	Outcome Assessment	Analysis Method	Transferability/ Generalizability	Other Confounding Factors	Indirect Comparison
England	√	√	NA	√	×	√	×	×
France	√	×	NA	×	×	×	×	√
Germany	√	×	NA	×	×	×	×	√
Scotland	√	×	NA	√	×	×	×	√

quality of life outcome, which were relevant for economic analysis. Regarding the uncertainties in the outcome assessment, SMC determined that there were limited data in some subgroups such as female, elderly, and more severely diseased patients. NICE noticed that from month 12 onwards, the Kaplan-Meier plots were heavily influenced by censoring data (that is, people who did not have an event during follow-up whose survival is unknown beyond the point at which they were censored), so very few patients remained at risk of mortality or disease progression. Therefore, the estimates of survival beyond 12 months were highly uncertain (Table 2.10).

2.2　SUMMARY OF STUDY LIMITATIONS

2.2.1　Comparative Studies

Most of the products investigated submitted only single-arm trials as pivotal studies, making the comprehensive analysis on the comparative effectiveness and comparative safety with alternative interventions to be challenging. The RCT study for Imlygic was determined to be associated with limitations in the comparator selection, which was not considered the SoC or the one currently applied in clinical practice.

2.2.2　Study Populations

The main issues with the study populations were indicated in the following aspects:

- The discrepancies between the patients recruited in the clinical trials and the label of the market authorization (Glybera, Provenge, Luxturna, and Zolgensma)
- The exclusion of patients with higher disease severity from the trial (Alofisel)
- The representatives of the study population to the whole population eligible for treatment (Zalmoxis, Yescarta, Kymriah, Tecartus, and Zynteglo)

2.2.3   STUDY ENDPOINTS

The main issues with the study endpoints were indicated in the following aspects:

- Patient clinical relevance and the validity of primary endpoints in a real-world setting (Glybera, Imlygic, Yescarta, and Luxturna), with most endpoints being surrogates
- Endpoint analysis on a post-hoc basis (Alofisel and Zalmoxis)
- The reliability of endpoints due to the intertest variability and difficulty in repeatability (Luxturna)

2.2.4   STATISTICAL ANALYSIS

The method for statistical analysis was criticized for two products, with limitations included:

- Incomplete information on the imputation procedure for missing values (Alofisel)
- Uncertainty in the estimation of the long-term survival using the Kaplan-Meier method (Tecartus and Kymriah indicated for DLBCL)
- Potential overestimation of treatment benefits resulting from full analysis set (FAS) analysis rather than ITT analysis (Kymriah indicated for both DLBCL and B-cell ALL)

2.2.5   CONFOUNDING FACTORS

The existence of confounding factors that biased the treatment benefits mainly included the following:

- The impact of dietary regimen on treatment effect (Glybera), because diet was not optimized before trial entry
- Waiting time and bridging treatment before availability of treatment (Yescarta and Kymriah)
- Differences in patient characteristics and withdrawal rates between two comparison groups (Imlygic and Provenge)
- The questionable appropriateness to synthesize the data from two heterogenous studies (Kymriah indicated for B-cell ALL)

2.2.6   DATA TRANSFERABILITY AND GENERALIZABILITY

The challenges in data transferability and generalizability were mainly around the following:

- The inconsistencies between the treatment pathways used in the study and clinical practice (Provenge and Kymriah indicated for B cell ALL)

- Representativeness of study populations for the whole population eligible for treatment (Alofisel, Yescarta, and Kymriah)
- The extrapolation of short-term evidence for long-term effects (Luxturna)

2.2.7 COMPARISON OF INDIRECT TREATMENTS

Comparisons of indirect treatments were submitted for Strimvelis, Imlygic, Yescarta, Kymriah, Zolgensma, and Tecartus, which were mainly derived using the data from systematic reviews and meta-analysis, retrospective observational studies, patient registries, or network databases. The main limitations of the indirect comparisons were:

- The heterogeneity in patient characteristics, differences in the outcomes investigated, potential bias due to uncontrolled confounding factors between two studies for indirect comparison, and the difficulties in drawing firm conclusions regarding relative effectiveness based on the poor evidence
- The inherent uncertainties for the studies used for indirect comparison included the small patient number, the incomplete information about the baseline patient characteristics, the non-traceability of retrospective studies, and the potential selection bias for studies included in the meta-analysis

2.3 DISCUSSION AND RECOMMENDATIONS

2.3.1 DISCREPANCY IN THE HEALTH TECHNOLOGY ASSESSMENT

Gene therapies have usually been approved based on immature data derived from non-comparative, open-labeled studies with short follow-up durations and small patient populations. Such limitations in study methodology have led to uncertainties regarding gene therapies' long-term efficacy and safety and have further made the precise assessments of treatment benefit and cost-effectiveness impossible (Table 2.11).

2.3.2 VALUE DETERMINATION

In general, value determination constituted one of the most important factors for the reimbursement decisions, while different countries showed varying perspectives on the weights allocated to each value element.

- France emphasized clinical effectiveness, disease severity, and rarity as well as the unmet medical needs of the disease area and patients, as can be seen in the fact that gene therapies (i.e., Kymriah, Yescarta, and Luxturna) were evaluated as having "important" actual clinical

TABLE 2.11

HTA Decisions for Approved Cell and Gene Therapies

	France	Germany	England	Scotland	United States
Glybera	Not recommended (SMR: insufficient)	Recommended (added benefit: proven; extent of added benefit: non-qualifiable)	NA	NA	NA
Imlygic	NA	Not recommended (added benefit: no; extent of added benefit: no)	Recommended with patient access scheme	Not recommended: in the absence of submission of MA holder	NA
Strimvelis	NA	NA	Recommended	NA	NA
Kymriah	Recommended for hospital use for both indications • B-cell ALL (SMR: important, ASMR: III); • DLBCL (SMR: important, ASMR: IV)	Recommended for both indications (added benefit: proven; extent of added benefit: non-unqualifiable)	Recommended for use within CDF for both indications, along with market access agreement	B-cell ALL: recommended with patient access scheme • DLBCL: recommended with patient access scheme	• At least a small net health benefit • ICER met the cost-effectiveness threshold
Yescarta	Recommended for hospital use for both indications (SMR: important, ASMR: III)	Recommended for both indications (added benefit: proven; extent of added benefit: non-qualifiable)	Recommended for use within CDF for both indications, along with market access agreement	Recommended with patient access scheme	• At least a small net health benefit • ICER met the cost-effectiveness threshold
Luxturna	Recommended for hospital use (SMR: important; ASMR: II)	Added benefit: proven; extent of added benefit: considerable	Recommended with commercial arrangements	Recommended with evidence development	• At least a small net health benefit • ICER higher than the cost-effectiveness threshold

TABLE 2.11 *(Continued)*
HTA Decisions for Approved Cell and Gene Therapies

	France	Germany	England	Scotland	United States
Zolgensma	Initially recommended with indication restriction SMA type I and type II with up to three *SMA2* genes (SMR: substantial, ASMR: III) -Re-evaluation: SMA type I and type II with one to two copies of *SMA2* genes (SMR: substantial, ASMR: III)	Evaluation suspended	Recommended with discounts or commercial arrangements	Recommended with patient access scheme	• High certainty that Zolgensma provides a substantial net health benefit for infantile-onset SMA • ICER higher than the cost-effectiveness threshold
Tecartus	Recommended for hospital use (SMR: substantial; ASMR: III)	Recommended (added benefit: proven; extent of added benefit: non-unqualifiable)	Recommended for use within CDF along with market access agreement	Recommended with patient access scheme	NA
Libmeldy	Recommended for asymptomatic children without clinical manifestation (SMR: important; ASMR: III); not recommended for symptomatic children with clinical manifestation	Recommended (added benefit: proven; extent of added benefit: substantial for children with late infantile (LI) or early juvenile (EJ) forms of MLD without clinical manifestations of the disease; non-unqualifiable for children with EJ forms of diseases with clinical manifestations	NA	NA	NA

(Continued)

TABLE 2.11 *(Continued)*
HTA Decisions for Approved Cell and Gene Therapies

	France	Germany	England	Scotland	United States
Zynteglo	Not recommended for TDT patients older than 35 years old; recommended for patients aged over 12 years to less than 35 years (SMR: important; ASMR: III)	Recommended (added benefit: proven; extent of added benefit: non-unqualifiable)	Evaluation suspended because the manufacturer will not commercialize the product in the UK	NA	NA

benefit. Moreover, this is also reflected in the fact that decisions for reimbursement will be highly dependent on the disease severity and unmet needs, in addition to the competence of clinical evidence. For example, Zynteglo, Zolgensma, and Libmeldy were only recommended in some subpopulations, not corresponding to the label of EU approval.

- Germany underlined the comparative benefits against available treatments. Therefore, the limitations of indirect historical comparisons were mentioned as one of the important reasons for the unavailability of accurate benefit assessments for Kymriah, Yescarta, Tecartus, and Libmeldy. Not surprisingly, Imlygic was evaluated with no added benefit due to the inappropriateness of comparators.
- In the UK, England and Scotland attached great importance to cost-effectiveness analysis. NICE defined a higher ICER threshold for ultra-orphan drugs evaluated under the highly specialized technology (HST) pathway. The SMC introduced a new approach for the assessment of ultra-orphan drugs, in which a higher uncertainty in economic analysis could be acceptable. For example, Strimvelis for ADA-related SCID and Tecartus for relapsed or refractory mantle cell lymphoma was recommended under the HST pathway in England, as was Kymriah for B-cell ALL treatment in Scotland.

2.3.3 Reimbursement and Affordability Strategies

Along with the disparity in value assessment, different strategies were adopted in each country to achieve prompt market access to innovative gene therapy without impairing healthcare affordability by reimbursing costly, yet effective, treatments.

- In England, all gene therapies, except Strimvelis, were recommended for use in combination with a commercial patient access scheme (PAS) to improve the cost-effectiveness profiles. Additionally, despite negative recommendations for routine use in England, three CAR-T cell therapies, Kymriah, Yescarta, and Tecartus, were accepted for interim use in the CDF during a further data collection period, after which, a reassessment will be scheduled to decide whether the products will be recommended for routine use.

- In France, all reimbursed gene therapies were restricted to be administered in qualified healthcare institutions. Moreover, the prescription decision for Luxturna must be examined by a multidiscipline expert panel, and the HAS will re-evaluate Luxturna's benefit after 5 years based on new evidence. The recommendations for reimbursement could be altered as more comprehensive evidence becomes available. For example, Zolgensma was evaluated to have "moderate" added benefits in patients with type I SMA and type II SMA with one to two copies of SMA2 genes in the reassessment, which is more restricted compared to the initial assessment, which covered a broader population with up to three copies of SMA2 genes.

- In Germany, the G-BA holds a conservative attitude in the assessment of the extent of added benefit, which is reflected in the fact that all gene therapies (except for Imlygic) were considered to have a "non-quantifiable" added benefit against the comparators. Additionally, all the assessments were valid for a limited time, with re-evaluation beginning after additional data collection is completed. This is due to the fact that G-BA put great emphasis on the comparative evidence against the appropriate comparators.

2.3.4 RECOMMENDATIONS FOR CLINICAL TRIALS FOR RMs

2.3.4.1 Study Populations
The recommendations for the selection of study populations of RM clinical trials are provided in Table 2.12.

2.3.4.2 Comparators
The recommendations for the selection of appropriate comparators of RM clinical trials are provided in Table 2.13.

2.3.4.3 Study Endpoints
The recommendations for the selection of relevant study endpoints of RM clinical trials are provided in Table 2.14.

2.3.4.4 Analysis Methods
The recommendations for the choice of an appropriate analysis method for RM clinical trials are provided in Table 2.15.

TABLE 2.12

Recommendations for the Study Populations of RM Clinical Trials

Study population	• Is the number of the study population large enough to examine the clinical difference? • Is the study population aligned with the label of market authorization? • Are patients with different severities included to represent diversity in the real-world setting? • Is the rationale for conducting the specific subpopulation analysis justified? • Are the baseline characteristics of the study population well recorded? • Are the important prognosis factors for disease well balanced between comparison groups? • Is it appropriate to combine all relevant clinical data as integrated evidence to expand the patient number size?

TABLE 2.13

Recommendations for the Comparators of RM Clinical Trials

Comparator	• Isa comparative study possible?
	   - Are other treatment options indicated for the same disease available?
	   - What is the standard of care for the management of this disease?
	 • Are the treatment regimens for the comparators applied in the study (such as the concomitant treatments and drug dosages) aligned with clinical practice in the real-world setting or MA label?

TABLE 2.14

Recommendations for the Study Endpoints of RM Clinical Trials

Study endpoint	• Were the validity and accuracy of surrogate endpoints for patient relevant endpoints sufficiently examined, such as systematic review? • Were the endpoints oversubjective and could be subject to assessment bias due to an open-label study design? • Was this endpoint reliable to investigate treatment effects, and was it repeatable irrespective of different investigators? • Was this endpoint pre-defined in the study protocol?

TABLE 2.15

Recommendations for the Analysis Method of RM Clinical Trials

Analysis method	• Are the analysis methods sufficient to capture the treatment benefits according to the size of the study population and follow-up duration? • Are the potential confounding factors (e.g., patient withdrawal from the study, the change of treatment, bridging treatment) fully considered? • Are the methods for the management of missing values reported?

TABLE 2.16
Recommendations for the Indirect Comparison of RM Clinical Trials

Indirect comparison	• What is the most suitable data source for indirect comparison: the natural history study, patient registry database, claim records, or well-conducted observational studies? • Are there any significant differences in the patient characteristics, treatment strategies, endpoint reports, follow-up durations and drop-out rates between study groups? • Are the important prognosis factors well considered and adjusted between studies with acceptable matching strategies? • Does the investigational study serving as a comparative group have good study quality, such as complete information on the patient baseline characteristics, sufficient patient numbers, and long enough follow-up duration? • Were the inclusion criteria of the meta-analysis fully explained? Was the risk of bias for studies included in the meta-analysis acceptable? • Was the use of indirect comparison pre-defined in the study protocol?

2.3.4.5 Indirect Comparison

The recommendations for the method of indirect comparison of RMs are provided in Table 2.16.

3 Challenges for Market Access of RMs

In the face of substantial uncertainties presented in the clinical evidence of RMs, decision-makers questioned extensively the durability of clinical benefits and the unforeseeable toxicities in the long run. Health technology assessment (HTA) bodies criticized the fact that limited evidence has deterred the robust assessments on the relative effectiveness and economic impacts of RMs (58). In particular, it is currently under extensive debates regarding whether the existing methodology for the value assessment and economic analysis is appropriate for RMs or whether a different framework that could better accommodate the specialty of RMs will be preferred. Regulators and HTA bodies therefore required that post-marketing studies be conducted to bridge the evidence gaps in the initial submission. However, previous investigations indicated that the post-marketing scientific obligations were generally poorly fulfilled in terms of meeting completion deadlines and disappointing study quality (25).

Moreover, despite no confirmatory evidence available to demonstrate the "curative" benefits of RMs, manufacturers charged excessively high prices with intentions to recoup the development investment (59). This made payers skeptical about the justification of the high prices of RMs in relation to the nontransparent disclosure of manufacturing costs and the paucity of clinical evidence (60). More importantly, the high up-front cost of RMs has threatened the sustainability of payers' affordability, considering that the existing payment system is more fitted for traditional medicines or biologicals that spread cost over multiple years.

The objective of this chapter is to describe the challenges faced by RMs regarding aspects of HTA evidence assessment, economic analysis, value assessment, price, affordability, and payment models. It also provides recommendations on the future efforts needed to navigate these challenges for safe and timely patient access of RMs.

3.1 CHALLENGES IN THE HEALTH TECHNOLOGY ASSESSMENT

3.1.1 UNCERTAINTIES IN THE CLINICAL EVIDENCE

- **Study design of the clinical trials**
  - **Small sample size**

Manufacturers may find it challenging, or even infeasible, to recruit enough patients for adequately sized trials in the cases of RMs targeting rare diseases (61). RMs were largely studied in small studies enrolling a limited number of patients; 47.2% of trials for RMs recruited fewer than 25 patients (38). Small trials are detrimental to the robust assessment of clinical and economic outcomes (62–65) owing to the poor representation of the enrolled population to the eligible population (66). In order to increase the sample size of clinical trials, researchers sometimes must compromise the homogeneity of the trials by including patients with different baseline characteristics (67). This would mean opening inclusion and exclusion criteria to broaden the population. This may widen the 95% confidence interval (distribution) of endpoint estimates in the trial, rendering it difficult or even impossible to achieve statistical differences in direct or indirect treatment comparisons. Additionally, for severe life-threatening diseases, patients may be reluctant to participate in a placebo-controlled trial, especially when RMs with the potential to cure are available (13). There is a significant risk of patients dropping out from trials when they will discover they are allocated to placebo or poorly effective therapy. Then patients may try to seek access to RMs through another way. Limited patient numbers and heterogeneity of patient populations raise questions about the transferability and generalizability of the results of clinical trials (67) (Figure 3.1).

- **Short-term data and extrapolation**

Although RMs are expected to have sustainable long-term treatment benefits, most clinical trials only provide short-term evidence, making the reliable assessment of the long-term outcomes usually impossible (68, 69) (Table 3.1). For example, the longest follow-up for axicabtagene-ciloleucel (axi-cel) was limited to a median of 15.4 months in the ZUMA-1 study, increasing the uncertainty around long-term overall survival (OS) and immunoglobulin usage (65). Although extrapolation of short-term evidence may provide a way of estimating long-term effects, it cannot replace good-quality clinical evidence (62). Additionally, the acceptance of the extrapolation method and the derived models vary across different HTA agencies (70). In some circumstances, extrapolation is just not feasible because of the absence of appropriate data; for example, the transition probabilities of rare diseases are unknown in most cases (71). It is also common to have a flat responder rate, even a 100% response rate, in gene replacement therapies. In that case there are no mathematical methods to extrapolate a flat rate of 100%. The only way is to build a scenario based on expert knowledge, which carries serious limitations in payers' eyes (71).

FIGURE 3.1   Gaps between efficacy and effectiveness.

TABLE 3.1

Design of Pivotal Clinical Trials for Cell and Gene Therapies

Brand Name	Design of Pivotal Studies	Study Duration, Patient Size	Primary Outcome (Surrogate Endpoint or PRO)	PRO as Secondary Outcome
Glybera	1) CT-AMT-011–01; 2) CT-AMT-011–02: Phase 2/3, multicenter, single-arm, open-label, dose-escalation study	1) 12 weeks, N = 14; 2) 18 weeks, N = 5	Reduction of fasting TG concentrations (surrogate endpoint): 7 out of 14 in CT-AMT-011–01 study; 2 out 5 in CT-AMT-011–02 study	Survival not collected; QoL at last follow-up not available
Imlygic	Study 005/05: Phase 3, multicenter, randomized, controlled, open-label study; comparator: GM-CSF	17.1 months, N = 436	Durable response rate (surrogate endpoint): 16.3% in Imlygic versus 2.1% in GM-CSF	No significant difference in OS (P = 0.0511); QoL not collected
Alofisel	ADMIRE-CD study: Phase 3, multicenter, randomized, double-blinded, placebo-controlled study	24 weeks (extension to 104 weeks), N = 208	Combined remission rate (surrogate endpoint): 49.5% in Alofisel versus 34.35% in placebo group	Survival not collected; no significant difference in the QoL
Zalmoxis	TK007 study: Phase 1/2, multicenter, single-arm, open-label, interventional studies	3 years, N = 30	Immune reconstitution (surrogate endpoint): 77%	OS for combined HSCT + Zalmoxis: 40% (1 year), 30% (2 years), 27% (5 years); QoL not collected
Strimvelis	AD1115611: Phase 1/2, multicenter, single-arm, open-label, interventional studies	3 years, N = 12	Survival rate (PRO): 100% in Strimvelis versus 67% in HSCT group	PRO as primary endpoint; QoL not collected

(Continued)

TABLE 3.1 *(Continued)*
Design of Pivotal Clinical Trials for Cell and Gene Therapies

Brand Name	Design of Pivotal Studies	Study Duration, Patient Size	Primary Outcome (Surrogate Endpoint or PRO)	PRO as Secondary Outcome
Provenge	IMPACT study: Phase 3, multicenter, randomized, double-blinded, placebo-control study	60 months, N = 512	Overall survival time (PRO): 20.6 months in Provenge versus 19.3 months in placebo	PRO as primary endpoint; QoL not collected
Luxturna	Study 301: Phase 3, multicenter, randomized, open-label, control study; comparator: SoC	1 year, N = 31	Multiluminance mobility testing (MLMT) (surrogate endpoint): 1.8 in Luxturna versus 0.2 in control	Survival and QoL were not collected
Yescarta	ZUMA-1 study: Phase 1/2, multicenter, single-arm, open-label study	12 months, N = 101	Complete response rate (surrogate endpoint): 47% in Yescarta versus 7% in the standard of care	OS: 59 (53%); QoL not reported
Kymriah for B-cell ALL	ELIANA study: Phase 2, multicenter, single-arm, open-label study	13.1 months, N = 92	Overall remission rate (surrogate endpoint): 82%	OS: 90.3% (month 6); 76.4% (month 12); clinically meaningful improvement in QoL among responders
Kymriah for DLBCL	JULIET study: Phase 2, multicenter, single-arm, open-label study	11.4 months, N = 147	Overall response rate (surrogate endpoint): 33.9%	Median OS not reached; small increase in QoL among responders
Zynteglo	1) Study HGB-204 and 2) Study HGB-205: Phase 1/2, multicenter, single-arm; open-label study; 3) Study HGB-207: Phase 3, multicenter, single-arm, open-label study	1) 32.1 months, N = 19; 2) 38.3 months, N = 4; 3) 5.6 months, N = 23	Transfusion independence (surrogate endpoint): 59.1%	No patient died; QoL not collected

TABLE 3.1 *(Continued)*
Design of Pivotal Clinical Trials for Cell and Gene Therapies

Brand Name	Design of Pivotal Studies	Study Duration, Patient Size	Primary Outcome (Surrogate Endpoint or PRO)	PRO as Secondary Outcome
Zolgensma	STR1VE study: Phase 3, multicenter, single-arm, open-label study compared with natural history study	24 months, N = 21	Survival rate (PRO): 67% in Zolgensma versus 25% in natural history control group	PRO as primary endpoint; QoL not collected
Libmeldy	Study 201222 was a non-randomized, open-label, prospective, comparative (non-concurrent control) study	8 years, N = 22	Gross Motor Function Measure score (GMFM): 72.5 versus 7.4 for late infantile MLD; 76.5 versus 36.6 for early juvenile MLD	All subjects with late infantile MLD remain alive (overall survival 100%); none of the pre-symptomatic EJ MLD patients treated died; 2/7 subjects in the symptomatic EJ MLD treated with Libmeldy died due to disease progression
Tecartus	ZUMA-2 study: Phase 2, multicenter, single-arm, open-label study	N = 130	Overall response rate (surrogate endpoint): 53/60 (88%: 77.4%, 95.2%); CR: 42/60 (70%: 56.8%, 81.2%)' PR (95% CI) 11/60 (18%)	Percentage of patients dead/alive: 21(28%) vs. 53 (72%)

- **Lack of appropriate comparators**

Conducting randomized controlled trials (RCTs) can be problematic for RMs, especially for rare diseases, because the available treatment is usually lacking or there is little knowledge of the disease management for the targeted population to inform the appropriate comparators or what is the standard of care (SOC) (72). Moreover, the selection of appropriate comparators can differ between countries depending on the clinical practice and product availability in different geographic areas (73). As illustrated in the case of talimogene laherparepvec (Imlygic: Amgen, USA) for metastatic melanoma, despite a positive reimbursement decision from the

England National Institute for Health and Care Excellence (NICE), it was rejected by the Germany Federal Joint Committee (G-BA) because the use of granulocyte-macrophage colony-stimulating factor as a comparator was not accepted (74).

- **Single-arm trial**

The rationales behind not conducting a comparative trial for cell and gene therapies included, but was not necessarily limited to, the argument that this type of study could increase irreversible damage to the patients and difficulty in recruitment from small patient groups (71). However, the single-arm nature of clinical trials (62, 75) limits all attempts at a robust assessment of relative benefits versus existing treatments, which constitutes one of the most important criteria in HTA decision-making. The lack of a randomized control arm represents a higher risk of bias and higher uncertainty around the actual effect size estimate (76). Although indirect comparisons (e.g., historical comparison or matched adjusted comparison) (Figure 3.2) could be applied to compensate for the uncertainties of non-comparative trials, inherent limitations, such as retrospective design and unmatched study populations, have undermined RM acceptance in the HTA decision-making (68, 77).

- **Blinding**

Furthermore, RMs may represent additional challenges for blinding clinical trials owing to their potential invasive method of administration (61). Under such circumstances, even if a comparative study was possible, fully blinded controlled trial designs may sometimes require potentially unethical sham procedures (61). For example, it would be difficult for an ethical committee to accept a double-blind study comparing nusinersen with repeated intrathecal administration vs. Zolgensma with a single IV administration. The rational to support a double-blind comparison may be seen as too invasive with little benefit from the scientific perspective for the targeted population.

- **Selection bias**

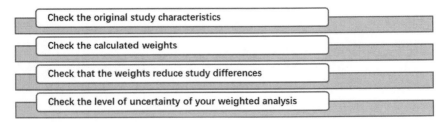

FIGURE 3.2   Summary of the steps for matched indirect comparison (MAIC).

Source: www.quantics.co.uk/blog/practical-steps-when-conducting-a-matching-adjusted-indirect-com parison-maic/

Evidence uncertainty may be further aggravated by selection biases. The lengthy process of RM production (78) may lead to patients being selected based on their general condition and ability to wait. In addition, patients with a better performance status (79) who are able to travel long distances to specialized cancer centers may be selected for clinical trials investigating RMs available in only one center (e.g., a single administration center in Italy for an autologous CD34+ enriched cell fraction that encodes for the human adenosine deaminase gene sequence [Strimvelis: Orchard Therapeutics, United Kingdom]) or a very few centers globally (79). Conversely, patients with highly aggressive diseases might not be able to receive RM treatment. Patient selection in the clinical trial may lead to overestimation of the treatment effect observed in pivotal trials (79).

- **Subgroup analysis**

Small studies make it more difficult to perform meaningful subgroup analysis to identify those patients with the highest possibilities to be responsive (61). This is critical when inclusion criteria of clinical trials vary widely for enrollment of a heterogeneous population (e.g., mix of children and adults) (38). For example, Roth et al. proposed that axi-cel may have different quality-of-life (QOL) outcomes in some clinical subgroup of relapsed/refractory diffuse large B-cell lymphoma (R/R LBCL), such as mediastinal large B-cell lymphoma. However, data limitations make outcomes in subgroup patients rather than overall R/R LBCL patients unable to be examined (65). In the economic evaluation of tumor-infiltrating lymphocytes for advanced melanoma, Retel et al. indicated that the patient subgroup that would have the best response to Tumor Infiltrating Lymphocytes (TIL) was unknown, and the cost-effectiveness outcomes may vary if appropriate biomarkers to identify the suitable patients are available (80).

- **Outcome of the clinical trials**
  - **Surrogate endpoints**

Surrogate endpoints were frequently used, whereas the absence of knowledge informing the disease characteristic raised questions about the validity and predictability of surrogate endpoints to clinically relevant endpoints (67). Explicit evidence is lacking to show how changes in surrogate endpoints will lead to the outcomes (e.g., QoL) that are critical for reimbursement decision-making (81). For example, alipogene tiparvovec (Glybera) was rejected by the French National Authority for Health (HAS), partly because of the uncertainty about translation of the primary endpoint (i.e., blood triglyceride level) to actual benefits (74).

- **Durability and "curative" effects**

Uncertainty surrounding the duration of treatment benefits was the major driver for the wide variation of incremental cost-effectiveness ratio (ICER) results for voretigene neparvovec-rzyl (VN) (82), valoctocogene roxaparvovec (83), and chimeric antigen receptor T-cell therapy (CAR-T) (68, 84–86). The clinical evidence available was insufficient in most cases to support the claims that RMs are "curative" (62). Even in some circumstances, the estimates for the sustainability of treatment benefits could be overly optimistic (72). The credibility of extrapolation based on standard modeling techniques (e.g., Markov models) for long-term outcomes (86, 87) was undermined because of the unreliability of input data. The cost-effectiveness analysis for RMs is highly sensitive to the long-term outcomes (e.g., remission rate and relapse rate), whereas a more conservative assumption may reverse the conclusions (86, 88).

- **Survival**

South et al. questioned the relative usefulness of OS as a primary outcome when only small trials are available because patients were counted as a treatment success despite having to receive alternative treatments. Under such circumstances, intervention-free survival may be a better indicator of effectiveness than OS (66). The use of mean OS over the median OS also triggered discussions. Walton et al. argued that the mean value did not really represent the life expectancy of patients, considering a small proportion of long-term survivors (maybe of a young age) would skew the mean life expectancy (72).

When follow-up is premature, a single "best-fitting" survival model may not be accurate to predict the survival distribution (89). Partitioned survival models often fail to properly incorporate the complexity of the disease and technology (90). Mathematical extrapolation may not be feasible when all treated patients experience 100% survival or response during the follow-up period. In that case, an alternative method for extrapolation relying on expert opinions may be most appropriate (71).

- **Additional benefits from a societal perspective**

Most cost-effectiveness analyses of RMs fail to take a whole-of-life view that takes the potential gains in non-health benefits into consideration. This is particularly relevant for RMs because a large percentage of RMs are indicated for childhood-onset, genetic diseases currently out of alternative treatments. If responsive to RMs, pediatric patients will have better education and employment opportunities in the future (88, 91, 92). The key reasons for neglecting broader societal values are the limited knowledge of both education and employment prospects of long-term survivors. Additionally, benefits in terms of alleviation of the burden of caregivers (physically and emotionally) are usually not accounted for (71). For example, the improvement in lifetime QoL resulting from remedying visual impairment after VN treatment, which was implied as 8% in the study by Zimmermann et al., could be significantly underestimated, particularly in light of feedback from patients, caregivers, and physicians (93), reflecting the neglect of additional values that will not adequately capture the clinical and economic outcomes of RMs. In the absence of methodological guidelines on how to include such non-health gains in economic evaluations, the inclusion of societal benefits of RMs in economic models often relies on simplistic assumptions and should be interpreted with caution (88).

- **Post-launch real-world evidence collection**

HTA agencies care about effectiveness for all the patients eligible for treatment in the real-world setting, not just efficacy in the selected narrow population included in the clinical trials (94). Especially for RMs, post-launch real-world evidence (RWE) collection has been increasingly emphasized by HTA bodies to compensate for the evidence limitations in the initial submission.

However, from the practical side, the infrastructures needed to easily capture RWE remain in development in most countries and will require additional investments (67). From the methodological side, the concept of RWE is immature, and despite several available guidelines, there currently exists no standard framework outlining requirements for RWE, such as how a specific item is defined and which type of design is required to address a specific question. Additionally, consensus on the minimal data sets required for post-market RWE collection is lacking (95). Existing databases may have limited abilities to capture the QOL, or more broadly, the patient reported outcome (PRO) needed for informing future reimbursement and pricing decisions post-launch (96).

In terms of RWE collection through patient registries, RMs centers reporting to the patient registry frequently rely on referring local physicians and hospitals, which are not directly connected to the registry infrastructure (95). Ideally, the patient registry is expected to track the continuous journey of patient care, not only in the specialist centers but also in the clinical data beyond hospitals where patients may need to travel. However, in most occasions, there is a lack of universal electronic medical records and compatibility between the various sources of medical records for the same patient to avoid important information slipping through the patient tracking network (97, 98).

When implementing the RWE, another important issue is the legitimacy of sharing patient data across different jurisdictions. Patient registries provide highly sensitive and private data, demanding a detailed comprehension of the legal implications for proprietary use of medical data (73). In certain jurisdictions, patient registries are becoming difficult to implement without an adjustment of legislation in place (97).

Finally, the multiplicity of registries for the same disease and different therapies creates an issue with comparability, double counting of patients when registries are aggregated based on published data, change in outcomes, and structure of registries avoiding the continuity of longitudinal data.

3.2   CHALLENGES IN ECONOMIC ANALYSIS

- **Limitations due to model input and structure**
  - **Resource use**

The estimation of total costs for RM treatment could be conservative because the follow-up duration is too short to investigate the long-term outcomes (80, 83). For example, proper modeling of CAR-T will require consideration of a wide

TABLE 3.2
Additional Costs Associated with CAR-T Cell Therapies

Process	Additional Costs on Top of Drug Price	Reference
Manufacturing and preparation	-Patient screening baseline assessment of lymphocyte subsets -Leukapheresis -Lymphodepletion therapy	Zhu et al. (102)
Administration	-Bridging therapy prior to administration -Combining chemotherapy before and after administration -Lengthy hospital stays -Intensive care unit stay	Champion et al and Prasad et al. (79, 103)
Post-administration (short term)	-Monitoring and management of adverse events (including severe cytokine release syndrome, CAR-T cell–related encephalopathy syndrome, pancytopenia, hypogammaglobulinemia, and infection) -For example, infusions of intravenous immune globulin for hypogammaglobulinemia; IL-6–receptor antibody tocilizumab for cytokine release syndrome	Roth et al. (104)
Post-administration (long term)	-The rates of treatment success vary: most patients will therefore still require other salvage therapies and palliative care -Frequent hospital visits for long-term follow-up	Champion et al. (103)

CAR: Chimeric antigen receptor.

breadth of resources that are necessary to provide the therapy (90). However, the long-term information on the incidence of treatment-related toxicities (e.g., B-cell aplasia and cytokine release syndrome [CRS]), for example, and the duration of treatment for the management of these toxicities (e.g., intravenous immunoglobulin) was impossible to collect in the clinical trials with a short follow-up duration (68, 75, 77, 84, 86, 99) (Table 3.2).

Long-term administrative costs for RM were not considered in most cases. For example, the cost of US Food and Drug Administration (FDA)–mandated 15-year monitoring for all patients receiving RMs was not included (91, 100). One special example is Strimvelis, which could only be used in one center in Milan, for which the traveling costs were omitted in the economic analysis (66). Moreover, payment mechanisms for RMs (e.g., bundled payment or fee-for-service) remained unknown; hence, potential mark-up costs were either excluded or relied on assumptions (99, 101).

- **Indirect costs**

Diseases targeted by RMs, which are usually severely debilitating, would be accompanied by substantial indirect costs related to special education (for children), social services, and reduced productivity (for adults) (71). However, such potential savings on indirect costs caused by RMs could be insufficiently considered, which is partly due to the absence of consensus on the preferable methods for estimating the indirect cost. For example, it is inconclusive how to decide between the human capital approach and friction cost approach when calculating the indirect cost owning to productivity loss (71). Additionally, gene therapies often provide an additional gene to replace a defective gene but do not impact the fundamental genetic information in germinal cells. As illustrated in the case of spinal muscle atrophy (SMA) type I, patients will potentially survive as a "normal" person but transmit the defective gene to the next generation. This may lead to an increase in the prevalence of the disease or require healthy embryo selection, although these economic costs are never considered (71). As such, patients may die (SMA type I) or may not be fertilized; by surviving they may contribute to increase the disease prevalence, especially for the most severe forms, if embryo selection is not implemented.

Alternatively, in the absence of data for indirect costs related to the diseases under investigation, indirect cost data from other diseases within the same therapeutic area may be used. For example, in the two cost-effectiveness analyses of VN for RPE65-mediated inherited retinal diseases (IRDs), indirect costs were sourced from populations with diabetic retinopathy (93) and populations with neovascular age-related macular degeneration (82). However, the indirect costs in these populations have little relevance to the patients with RPE65-mediated IRD, who are typically young, and their educational options became increasingly restricted as visual impairment progressed, leading to lower incomes and long-term caregiver burden. As a result, the indirect costs associated with the disease, and the potential economic benefits of VN that are generated through indirect cost reduction, could be significantly underestimated.

- **Discount rates**

RMs are likely to involve high up-front costs occurring years before the health benefits could emerge, suggesting that a discount rate will have a significant impact on the ICER estimates (71). This generated debates on whether specific discount rules, such as a lower discount rate than conventional drugs, or a differential discount rate for health outcomes and for costs, should be applied for RMs to better accommodate the long-term benefits (76). The NICE assessment framework allows a non-reference case discount rate of 1.5% to be applied when:

- Treatment restores people who would otherwise die or have a very severely impaired life to full or near-full health

- When this is sustained over a time period of at least 30 years
- The committee is satisfied that the health effects are highly likely to be achieved and the introduction of the technology does not commit the National Health Service (NHS) to significant irrecoverable costs

However, it seems challenging to make the rationale for the 1.5% discount rate convincing with a lack of robust data supporting curative claims of RMs (105). The 2016 NICE report for CAR-T cell therapy identified that discounting the rate had a very significant impact on the analyses; however, this is because these are one-off technologies with a high up-front cost and not due to curative potential. The report concluded that the current NICE appraisal methods and decision framework are applicable to RMs and cell therapies. Likewise, the USA Institute for Clinical and Economic Review (USA ICER) concludes that it is not compelling to use a unique discount rate rule for RMs (106).

- **Time horizon**

Economic analysis using a lifetime horizon would be favored by HTA bodies given the presumably "curative" nature of RMs. However, there are no clinical trials with follow-up duration over a lifetime available to enable such economic analysis. In the case of using short-term clinical evidence of RMs to inform economic analysis of the lifetime horizon, it may lead to a wide variation of ICER because of the substantial uncertainties in the long-term clinical benefit or survival (67, 71).

3.2.1 CHALLENGES IN VALUE ASSESSMENT

- **Method for utility elicitation**

The generic instruments commonly used for utility estimates may have insufficient ability to evaluate the overall QOL outcome. For example, the EQ-5D series emphasizes the importance of clinical benefits but may neglect the occurrence of adverse effects (107). Besides, utility data are often collected from small patient groups included in the clinical trials of RMs (108). Furthermore, generic measurements (e.g., EQ-5D) may not be appropriate for assessing health status in very young children, who are generally incapable of correctly describing their experiences (71). Concerns are raised regarding whether it is acceptable to rely on the proxy report of QOL outcomes (parents or caregivers) and what is the best approach to minimize the potential investigator bias (71, 108).

If that patient-level utility data from clinical trials were lacking, vignette-based studies could be used to directly elicit the utility data (71). In this approach, detailed descriptions of each health state (vignette) are developed using different sources of information (e.g., patient and physician interviews); then the general public are asked to rate these states in a stated-preference experiment (e.g., time trade-off or standard gamble) (109). Although vignette studies may be acceptable

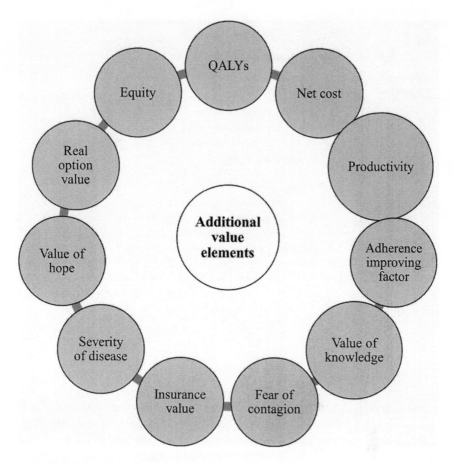

FIGURE 3.3 Additional elements of value.

Source: Figure adapted from Lakdawalla DN et al. Defining Elements of Value in Health Care—A Health Economics Approach: An ISPOR Special Task Force Report. License number for reuse: 4970110719583

to approximate a societal viewpoint, this method is limited by the validity of the vignette descriptions because 1) vignettes are not able to fully reflect the varied health-related QOL (HRQOL) experience among patients within a given vignette health state, and 2) there is a risk of investigator bias in developing vignettes if physicians or caregivers are surveyed (109).

Additionally, utility is measured using probabilities as assessment tools, and respondents may not have a good understanding of the concept of probabilities (110). Alternatively, utility elicitation also resorts to assumptions or extrapolation (80, 82, 88), leading to the biased estimates of utility. For example, in the cost-effectiveness study for VN, utility values from patients with diabetic retinopathy were used for patients with RPE65-mediated retinal disease (87, 93) (Figure 3.3; Table 3.3).

TABLE 3.3
Broad, Societal Value Elements Associated with RMs

Value Elements	Details for Additional Values Brought by RMs
Labor productivity	• Rare, genetic diseases may have disease onset in the newborn or childhood period; thus "curative" RMs could increase their future employment opportunities and productivity in the workplace.
Adherence-improving factors	• One-off treatments protect patients from frequent hospital visits to begin or maintain treatment.
Severity of diseases	• RMs potentially provide a "curative" treatment for life-threatening diseases or individuals near the end of life.
Value of hope	• Severely ill patients may be willing to trade off some survival (e.g., undertaking a risky procedure) for a chance at a "cure."
Equity	• RMs may bring value to equity issues by targeting rare diseases with high unmet needs; thus those unlucky patients previously suffering from untreatable diseases now become treatable.
Scientific spillover	• RMs, through their novel mechanism of action, may unlock the potential for future generations and stimulate the future discovery efforts by improving knowledge about the underlying causes of diseases.

- QALY calculation

Cost-effectiveness analyses assume a quasi-egalitarian value judgment that a quality-adjusted life-year (QALY) is equally weighted across individuals (e.g., children or adults) and conditions (e.g., end of life or less severe disease). Such an approach has been scrutinized for ignoring the fact that QALY improvement in certain diseases may deserve more appreciation (108).

The method for QALY calculation is another source of criticism (73). It is calculated as a sum of utility and duration of each health state, assuming mutual independence between utility and duration of health states. However, this assumption may not always hold, as illustrated in the SMA case, in which the health state of "inability to stand" will not have the same utility for infants younger than 6 months as for children up to 3 years of age (71). Moreover, as elaborated previously, unreliable utility data significantly undermine the robustness of QALY calculations (62, 105).

- ICER

A fixed threshold of ICER does not account for the fact that payers may value broad societal benefits that are not captured in the QALY when making reimbursement decisions (76). Moreover, what counts as an appropriate threshold for ICER seems arbitrary and does not necessarily facilitate justified resource distribution (111). It was proposed that the ICER threshold can be increased with more flexibility in the case of RMs characterized by innovativeness and extensive social values (112). For example, in England, the ICER threshold was raised to £100,000 (instead of £20,000 to £30,000) for highly specialized technologies (113). Such an approach might be meaningful to appropriately reward value created by RMs and to incentivize future investments (114). However, it must also be recognized that we are operating under a fundamentally flawed model of price setting; in some circumstances, ICER calculations can not only determine the cost-effectiveness but also how drug manufacturers may artificially inflate prices (111). Future research will be needed to explore whether a higher ICER threshold should be applied to RMs

- **Preference for one-time therapy over long-term administration of drugs**

It must be acknowledged that the presence of substantial benefits (i.e., non-health gains and reduction in indirect costs) associated with RMs is only possible when the assumption that the RMs are "curative" really holds. However, taking one step back, there are controversies regarding the level of certainties for this assumption. For example, in the case of retinitis pigmentosa, public expectations for a cure would be a therapy that improves vision from complete blindness to full sight, although the individuals eligible for RMs were not fully blind, nor did they restore full sight after treatment (87). From an economic perspective, a "cure" could be defined as eliminating the need for additional downstream costs for disease treatment or monitoring. Based on this definition, RMs (e.g., CAR-T) showing transient benefits and requiring substantial complementary treatments could not be qualified as "curative" therapies (115, 116).

Empirical evidence is inconclusive whether one-off RMs would be valued more highly by society than treatments that offer the same "total" health gains through marginal gains over many years (76, 116). These uncertainties make it clear that HTA bodies cannot make a straightforward assessment on the value of RMs simply because of their "innovative" or "transformative" characteristic; it is debatable whether innovation itself should be rewarded financially (97, 117).

- **How to perform value assessment systematically**

Although economists argue for a societal perspective when looking at costs and benefits, in reality, HTA agencies tend to take a narrower perspective (118). This is partly because of the absence of a protocol standardizing the methods of assessing social and ethical implications of health technology (73, 111). HTA guidelines are silent about the role of empirical studies for assessing the value of specific products from the perspective of patients or the general public (76). Consequently, the assessment of additional value elements is often unstructured and less transparent (67). One example is NICE's end-of-life criteria, which were difficult to interpret, indicating further clarifications were necessary to ensure consistency in decision-making (72). There are also risks of double counting additional value elements in QALY calculation and "other benefits" that are already included as part of most HTA reports (106).

Furthermore, the absence of consensus on the standard approaches for the value assessment of RMs will lead to heterogeneity in coverage recommendations across different healthcare settings, depending on how HTA agencies balance the potential clinical benefits and the underlying uncertainties (119). Such divergence is perhaps more visible when considering RMs than conventional therapies owing to their complexity and high up-front cost (73). Multicriteria decision analysis (Table 3.4) shows promise but has had limited application in the decision-making process because it is burdensome to implement, ignores opportunity cost, and requires more research to determine how to incorporate many value elements into a single value output (67, 116).

TABLE 3.4
Steps in Multiple-Criteria Decision-Making

Step 1: Selecting and structuring criteria

- Identify criteria relevant for evaluating alternatives

Step 2: Measuring performance

- Gather data about the alternatives' performance on the criteria and summarize this in a "performance matrix"

Step 3: Scoring alternatives

- Elicit stakeholders' preferences for changes within criteria

Step 4: Weighting criteria

- Elicit stakeholders' preferences between criteria

Step 5: Calculating aggregate scores

- Use the alternatives' scores on the criteria and the weights for the criteria to get a "total value" by which the alternatives are ranked

Step 6: Dealing with uncertainty

- Perform uncertainty analysis to understand the level of robustness of the MCDA results

Step 7: Reporting and examination of findings

- Interpret the MCDA outputs, including uncertainty analysis, to support decision-making

Source: Thokala P et al. Multiple Criteria Decision Analysis for Health Care Decision Making—An Introduction: Report 1 of the ISPOR MCDA Emerging Good Practices Task Force. License number for reuse: 4970370033059

3.3   CHALLENGES IN PRICING, REIMBURSEMENT, AND AFFORDABILITY

3.3.1   CONTROVERSY SURROUNDING THE HIGH PRICE OF RMs

It should be recognized that, in terms of pharmaceutical economics, being "cost-effective" differs widely from being "affordable" (120) (Figure 3.4). A society may be comfortable with absorbing higher costs for lifesaving products, but where the boundary between acceptable and unacceptable price should be drawn is much more complicated (111). The fundamental question is, how high a price should be charged for extension of life (121)?

- **Lack of transparency**

Pricing policy, especially in the United States, represents a complex and politically fraught subject, yet transparency of price-setting policies will become increasingly imperative with the arrival of expensive RMs (88). Manufacturers claim that there are additional costs driven by the personalized nature of RMs, which means that each product demands an individual process to manufacture and deliver it (Table 3.5). However, the research and development (R&D) costs of bringing an RM to market might have been shared between public and private players; therefore, public expectation of a lower price seems reasonable (79). Nevertheless, the fact that CAR-T was charged approximately 24 times higher than the manufacturing cost has fueled the wide debates over its fairness (79). Although pricing may be calculated to reward societal values and indirect health benefits, as discussed previously, how to quantify these additional values into HTA and price-setting is far from transparent (119).

- **Price of the alternative treatments**

Traditional value-based pricing based on a cost-effectiveness analysis will remain unjustified if overpriced comparators are used as price benchmarks. For example, nusinersen (Spinraza) was selected as the comparator for onasemnogene abeparvovec (Zolgensma) when performing the cost-effectiveness analysis and setting the price, and the price of Spinraza was arguably overpriced (59). Conversely, if comparators with cheap prices were selected, the cost-effectiveness and price

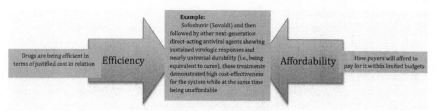

FIGURE 3.4   Gaps between efficiency and affordability.

TABLE 3.5

Prices of Cell and Gene Therapies Approved in the European Union

Brand Name	Manufacturer	Authorization Date	Market Withdrawal	Price at Market Entry US$
Gene therapy				
Glybera	UniQure	10/25/2012	10/28/2017	$1,206,751
Strimvelis	GlaxoSmithKline	5/26/2016	Marketed	$738,223
Imlygic	Amgen	12/16/2015	Marketed	$357,309
Kymriah	Novartis	8/22/2018	Marketed	$441,538
Luxturna	Spark Therapeutics	11/23/2018	Marketed	$425,000
Yescarta	Gilead/Kite	23/08/2018	Marketed	$373,000
Zynteglo	Bluebird Bio	29/05/2019	Marketed	€1.575,000
Zolgensma	Novartis	18/05/2020	Marketed	$2,125,000
Tecartus	Gilead/Kite	14/12/2020	Marketed	$373,000
Libmeldy	Orchard	17/12/2020	Marketed	$3.900,000
Abecma	BMS + Bluebird	18/08/2021	Marketed	$437,968
Breyanzi	BMS	27/01/2022	Marketed	$410,300
Tissue-engineered products				
ChondroCelect	TiGenix	11/16/2009	7/29/2016	$21,926
MACI	Vericel	6/27/2013	09/05/2014	$21,926
Holoclar	Chiesi Farmaceutici	2/17/2015	Marketed	$93,432
Spherox	Don AG	07/10/2017	Marketed	$18,950
Cell therapy				
Zalmoxis	MolMed Spa	8/18/2016	Marketed	$814,780
Provenge	Dendreon	09/06/2013	05/06/2015	$110,920

of RMs will be compromised, and the reimbursement and the patients' access to RMs will be hampered because developers may decide to forego the market access decisions (122).

- **Pathway for commercialization and its impact on price**

The cost of R&D may not be the only determining factor for the price-setting of RMs. The commercialization pathways of RMs also play a big role: if the manufacturers of RMs already have traditional drugs for the same indications within their product portfolio, their incentives to offer RMs for a low price will decrease, because the RMs would disrupt the profit of their existing market (92). The possibility that future expenses for RMs may increase as a result of the change of the product positions in the treatment protocol cannot be ruled out (123). For example, the combination of CAR-T and immune checkpoint inhibitors may result in improved clinical outcomes, increasing the chance that RMs are used earlier, even as first-line therapy (77). However, it seems less surprising that additional treatment benefits will be offset by a higher price tag (91).

3.3.2   High Up-Front Costs Threaten the Existing Payment System

3.3.2.1   Unique Challenges Due to the One-Off Treatment Nature

RMs will cause substantial up-front costs through single administration (97), so that it does not "neatly" align with existing payment models designed for small molecules or biologics administered on a regular basis (117) (Table 3.6). The

---

TABLE 3.6
Comparison between RMs and Traditional Medicines

	Regenerative Medicine	Traditional Medicines
Mode of action	• One-off treatment with high up-front cost	• Treatment on regular basis with treatment cost spread over many years
In case of ineffectiveness	• Unable to withdraw treatment and payment even if proven to be ineffective or unsafe	• Able to discontinue treatment and payment if proven to be ineffective or unsafe
Competition from generic	• Difficult to employ competition from generic or biosimilar products to drive the price down because of the challenges in the manufacturing and ethical issues of original RM products	• Reduction in drug price is expected when generic products or a biosimilar arrives after patent expiry
Targeted disease	• Difficult to employ competition between two brand RMs to reduce the price, as the prevalent patients will be largely targeted with the first RMs, while leaving only a small proportion of incident patients to be treated	• The market share for the first traditional medicines could be taken by the following products due to the diverse treatment benefits, and non-rare diseases with large patient populations leave more room for competition
Following treatment	• Patients already receiving gene therapies will be precluded from the further gene therapies due to the immune intolerance to the viral vectors, indicating the loss of "real option value" for future, better treatment options	• Patients can switch to future treatment options if available; the "real option value" was presented
Additional cost	• The additional administrative costs, such as personnel training, genetic testing, and patient travel to qualified medical centers, were unable to be sufficiently quantified; thus, it is difficult to make accurate estimation of overall cost	• Additional cost may be relatively low; treatment-related costs were well captured in the in-patient and out-patient electronic health records
DRG coverage	• DRG may not sufficiently cover the overall cost related to the administration of RMs (e.g., management of toxicity), placing the hospital in a profit-loss position	• DRG encourages the optimized use of high-cost drugs while maintaining (improving) the treatment outcomes and reducing the total cost

greater financial risk associated with RMs is that there is no opportunity to discontinue treatment or withdraw payment if they are proven to be ineffective or unsafe in the post-launch (118). Although the near-term cost burden is likely to be relatively small given that existing and emerging RMs are mostly indicated for rare diseases, it is the cumulative effects of multiple RMs targeting more prevalent diseases that would cause financial unsustainability in the longer term (124).

Payers will be facing an opportunity cost dilemma. Nisham et al. indicated that the opportunity cost for 200 patients treated with a single dose of CAR-T is around £56 million in England, which would otherwise enable healthcare systems to treat around 630 patients for lung cancer and 4,435 patients undergoing coronary artery bypass grafting to treat cardiovascular disease (125).

3.3.2.2 Downstream Costs to Be Considered

It is noteworthy that the drug price of RMs is just the "tip of the iceberg" when the additional administrative burdens are also factored into the total cost calculation (79). For example, in the case of CAR-T, other costs for managing hypogammaglobulinemia (65) and CRS (79) cannot be marginal. Moreover, in addition to the bridging treatments needed before (or even between) CAR-T treatments, some patients may still require salvage therapies and palliative care after the administration of CAR-T because of variations in patient response (123). Payers mentioned equal importance of clinical and financial aspects of products in the decision-making process. The products with good value for the money represent avoidance of downstream costs for disease management. Once more, this raised arguments that high-cost RMs, albeit effective, will greatly increase healthcare budgets if they cannot reduce the need for other high-cost therapies and downstream costs (91, 115).

3.3.2.3 Free-Rider Issues

Rose et al. indicated that the reimbursement of RMs is sometimes difficult to justify if payers will not capture all economic benefits from their payment activity (126). One example is RMs for wound care, for which treatment costs are met by acute care providers, yet they likely offer a greater savings to social care providers because of the decrease in bandage usage and home visits (126). More imperative, the high up-front cost of RMs creates challenges for health systems with multiple payers (e.g., in Germany and the United States) and in systems in which coverage decisions are made by subnational jurisdictions (e.g., in Sweden and Canada). Free-rider issues, in which later payers benefit from the payment action of the former payers, could happen in the situation of enrollment churn (patients switch insurance plans) (127). This significantly impairs payers' willingness to endorse RMs. Theoretically, the longer it takes for the benefits to materialize, the more incentives insurers have to reject patients in need of RM treatment (127).

3.3.2.4 Consequence of Insufficient Coverage of RMs

- **Inequity in patient access**

Restrictions in the coverage of RMs would potentially lead to unequal patient access to RMs. Currently, the logistics of giving CAR-T require the patient to be sick enough but not overly sick to be eligible for therapy. Given these characteristics, some patients may be clinically stable to wait long enough for approval at the beginning but quickly deteriorate while awaiting approval (121). The availability of RMs is only restricted to a limited number of qualified centers, which may also have important equality considerations because many people have no access to major treatment centers in their place of residence (128). Silbert et al. discussed that complicated issues, including patient age, insurance coverage, and clinician bias, might have a perverse impact on patient access to treatment (111).

- **Obstacles in the clinical adoption due to lack of funding**

In the United States, for drugs used on an inpatient basis, all services, including drug expenditures, will be subject to diagnostic-related group (DRG) rules (67). However, technology fitting into such a bundled payment model may not enable separate payment for drugs from the procedure, such as cell therapies used as an adjunct to hematopoietic cell transplantation (117). Additionally, the attribute of DRG-based reimbursement creates a scenario whereby the cost of RMs cannot be fully covered. Although such disconnection between the cost of innovative therapies and existing DRG-based payment is not uncommon, the efforts of seeking higher payment to accommodate a more costly therapy are time-intensive and represent only a temporary fix in most cases (129). As a result, inpatient administration of RMs may be financially infeasible for hospitals, even with the policy of outlier payment and a 65% (up to $242,450) new technology add-on payment. For example, hospitals may lose up to $200,000 for each CAR-T administered, which discourages hospitals from providing it (84).

3.4 CHALLENGES IN INNOVATIVE PAYMENT STRATEGIES

3.4.1 OUTCOME-BASED PAYMENT

- **Lacking appropriate outcomes**

An outcome-based payment mechanism requires clinically relevant and measurable outcomes, which are lacking for some RMs targeting rare diseases (67). Moreover, the measurement of benefits cannot be straightforward; for example, where patients have comorbidities, the outcome assessment may be more complex than a simple "yes" or "no" answer (76). On the other hand, outcome-based payment has limited capacity to compensate for adverse effects. Although financial adjustments can be made after RMs are proven to be ineffective, harm to patients, such as sudden death or severe adverse events, may be irreversible (62).

Furthermore, time for outcome assessment is critical for the successful implementation given the substantial time lag (sometimes years) between administration and the evident clinical benefit or harm (97). For example, Novartis proposed an outcome-based arrangement for tisagenlecleucel, in which it would charge the U.S. Centers for Medicare & Medicaid Services only if patients achieved remission by the end of the first month (86). However, because tisagenlecleucel' s high initial remission rates are also followed by loss of response and relapse (28.9% of patients who achieved an initial response reported relapse prior to the last visit), this model based on outcomes at early time points may not materially mitigate the risk compared with a traditional payment model (86). Payment arrangement for tisagenlecleucel in Italy, where payments will be made in stages upon enrollment and after 6 months and 12 months of treatment, could be more meaningful to reduce clinical uncertainties (96).

- **Administrative and legislative barriers**

Apparently, not all cell and gene therapies are eligible for outcome-based payments. For example, Skysona, one gene therapy indicated for the treatment of the neurological disorder cerebral adrenoleukodystrophy (CALD), received market authorization from the FDA in October 2022. Bluebird, the manufacturer of Skysona, is planning to charge $3 million per treatment, making it the most expensive drug in the world so far. The company claims the price of Skysona reflects the substantial benefits that the drug provided to slow down the progression of neurological dysfunction of patients suffering from irreversible and fatal rare diseases. However, Bluebird suggested that an outcome-based payment is challenging to implement for both the manufacturer and payers because of the rarity and complexity of CALD.

Moreover, the key to successful implementation of outcome-based payment is the careful collection of relevant and unbiased data (76), which is associated with significant administrative burden (73). Kefalas et al. estimated that, over a 10-year period, implementing the market entry agreement of hypothetical CAR-T for 50 new patients would require an additional 6,401 staff days, with the forecasted financial burden being £871,707 (78).

There are also regulatory barriers to implementing these contracts in the United States, such as the Medicaid Best Price, the federal Anti-Kickback Statute, and regulations on patient confidentiality (67). Finally, outcome-based payment provides no solutions to address enrollee churn when patients leave for other insurers (130).

3.4.2 Annuity Payment/Installment Payment

Some argue that the concept of a capped annuity is theoretically attractive but impossible to implement in most healthcare systems (120). Without being combined with outcome-based elements, this approach does not address uncertainty regarding actual drug performance (130). However, installment payments linked to clinical outcomes will necessitate clearly defined milestones (67), for which it is sometimes challenging to reach agreements between payers and manufacturers regarding what constitutes "success" and "failure" of treatment (76). Furthermore, enrollment churn remains unsolved because the former insurer is responsible for the loan payments even after enrollees switch insurance plans.

As possible solutions, annuity payments should spread over the duration of benefit and be paid continuously by the current plan which has the enrolled (previously treated) patients (115). However, it should be acknowledged that transferring annuity contracts from one plan to another without associated provision of services is currently not possible. Payers with previous annuity obligations would need to provide cumbersome claims data (120). Later plans might refuse patients, or companies with self-funded health insurance could resist hiring new employees with ongoing annuity payments, leading to discrimination against patients (120).

3.5 DISCUSSION AND RECOMMENDATIONS

3.5.1 SUMMARY OF THE STUDY FINDINGS

In general, the current HTA system may place RMs at a disadvantage (131) considering the HTA system is not designed to fully accommodate the specialty of RMs (112). Manufacturers have not proposed solutions to address the evidence limitation of RM at the time of launch. The substantial uncertainties in the comparative effectiveness and durability of clinical benefits constituted the biggest challenge for the HTA of RM. Furthermore, limitations in clinical evidence undermine the robustness of the economic analysis, which must rely on intensive extrapolations and assumptions on the long-term outcomes. Additionally, the model inputs for cost and utility were obtained from heterogeneous sources instead of reliable clinical data, and the additional administration costs associated with RMs were largely omitted from the cost estimates.

With regard to the value assessment of RMs, it is still inconclusive whether the innovative nature should be rewarded and whether the one-off RMs would be valued higher than chronic treatments from the perspective of patients. If RMs provide significant therapeutical benefit, their indirect benefits would also be substantial for both the patient and caregiver. However, the biggest difficulties are in incorporating these broader value elements into the HTA decision-making process in a systematic way (Table 3.7).

TABLE 3.7

Challenges in the Health Technology Assessment of Regenerative Medicines

	Limitations of Clinical Trials Increased the Uncertainties in the Clinical Evidence	Reference
Study design	RMs targeting rare diseases were investigated in clinical trials with a small patient sample size. To expand the sample size, patients with heterogeneous characteristics were enrolled.	Hanna (10); Drummond (14)
	Traditional RCTs are difficult to conduct due to lack of appropriate comparators; the invasive method of administration made the blinded, placebo-controlled trials unethical.	Hampson (9)

(Continued)

TABLE 3.7 *(Continued)*
Challenges in the Health Technology Assessment of Regenerative Medicines

	Limitations of Clinical Trials Increased the Uncertainties in the Clinical Evidence	Reference
	Single-arm trials limited the assessment of the relative effectiveness and made the effect size estimates uncertain.	Walton (26)
	Indirect comparison was scrutinized about the comparability of the study population and unadjusted confounding factors.	
	Clinical trials of short follow-up duration cannot give a robust answer about the "curative" benefits of RMs in the long run. Extrapolation sometimes was hindered because of the absence of reliable model inputs.	Roth (16)
	Lengthy manufacturing processes of RMs and accessibility of RMs in restricted medical centers potentially caused patient selection bias, which could result in overestimation of treatment effects observed in pivotal trials.	Prasad (33)
	Subgroup analysis to identify the patient group with the highest chance to be responsive was challenging.	Retel (48)
Clinical Outcomes	The validity of correlations between surrogate endpoints and clinical relevant endpoints was uncompelling due to the lack of knowledge on the disease characteristic to inform their correlation.	Qiu (28)
	The durability of treatment benefit was the major driver for the wide variation of ICER outcomes for RMs.	Johnson (35)
	Survival data were derived from short-term clinical evidence. There were controversies about the selection of survival endpoints (e.g., OS versus IFS; mean OS versus median OS) and the methods for modeling the long-term survival.	South (15); Walton (26)
	Potential gains in non-health benefits to patients, caregivers, and all of society brought by RMs could be substantial, but were barely considered in the value assessment.	Aballéa (22)
Real-world Evidence	Constricted financial resources of manufacturers limited their capacities to build up a data infrastructure to collect a large amount of high-quality data, which was further exacerbated by the complex logistic issues of RMs.	AMCP Forum (44)
	There is currently a lack of a standard framework specifying requirements for RWE, such as how to define the items to be collected and the preferable method for evidence collection.	McGrath (45)

TABLE 3.7 *(Continued)*
Challenges in the Health Technology Assessment of Regenerative Medicines

	Limitations of Clinical Trials Increased the Uncertainties in the Clinical Evidence Reference	
	A lack of universal electronic medical records disrupted the continuity of evidence generation through patient registries when patients travel to other regions.	Carr (47)
	Ethical and legislative challenges are present in sharing confidential patient data across agencies and countries.	Goncalves (27)
Challenges in the Economic Analysis of RMs		
Model input	The wide breadth of resources for administration of RMs was omitted, such as the cost for the management of adverse events, the companion genomic testing, long-term patient monitoring, and potential provider mark-ups.	Raymakers (41)
	It was challenging to quantify the indirect cost of the diseases treated by RMs due to the data scarcity and lack of consensus on the preferable method for indirect cost estimation (e.g., human capital approach or friction cost approach).	Coyle (8); Aballéa (22)
	Controversies regarding whether specific discounting rules for costs and benefits should be applied for RMs while using a lower discount rate (e.g., 1.5% or 0%) was less convincing in the lack of robust data supporting curative claims.	Angelis (53); Jorgensen (54)
	Although economic analysis using a lifetime horizon is preferred by HTA bodies, the clinical trials of short-term follow-up duration have limited capacities to support such exercises, leading to significant variabilities around the ICER.	Drummond (14)
Value assessment	Utility data were derived from heterogeneous resources or relied on assumptions.	Lloyd-Williams (11)
	Generic instruments (e.g., EQ-5D) may have insufficient ability to evaluate the overall QOL outcome, especially for young children. The vignette-based method (e.g., time trade-off) is limited by the validity of vignettes that are developed.	Wolowacz (60)
	Unreliable utilities data undermine the robustness of QALY calculations. Additionally, a quasi-egalitarian value judgment based on QALY ignores the possibility that QALY improvement in severe diseases may deserve more appreciation.	Goncalves (27)

(Continued)

TABLE 3.7 *(Continued)*

Challenges in the Health Technology Assessment of Regenerative Medicines

Limitations of Clinical Trials Increased the Uncertainties in the Clinical Evidence	Reference
The fixed threshold of ICER does not account for the possibility that decision-makers may consider broader values that are not captured in the QALY calculation.	Johnson (35); Pearson (66)
No consistent evidence is available to suggest whether one-off RMs are valued more highly by society over treatments that offer the same "total" health gains through marginal gains over many years.	Husereau (58); Pearson (66)
There is no standard definition or benchmark to determine whether the RMs are "innovative" and "transformative" enough, or if they merit a higher value by their innovative nature.	Faulkne (69)
The assessment of the social and ethical implications of health technology is often less transparent. Multicriteria decision analysis has limited applications due to the administrative burdens and the lack of standardized methodology protocol.	Husereau (58); Angelis (53)

EQ-5D: EuroQoL-5 Dimension; HTA: Health Technology Assessment; ICER: Incremental Cost-Effectiveness Ratio; IFS: Illness Free Survival; OS: Overall Survival; QALY: Quality-Adjusted Life-Year; QOL: Quality of Life; RCT: Randomized Controlled Trial; RM: Regenerative Medicine; RWE: Real-World Evidence.

The high prices of RMs raised controversies regarding whether such prices are justified in relation to the manufacturing cost and clinical benefits. Payers will face a financial crisis to cover upcoming RMs targeting not only rare diseases but also more prevalent diseases. The constricted payers' budget and divergence in the coverage policies will possibly lead to unequal access to RMs.

Outcome-based agreements are increasingly used for RMs by payers to share the risk and alleviate the short-term financial burdens. However, there are practical and legislative hurdles to overcome to allow the broader use of these agreements (Table 3.8).

3.5.2 Future Recommendations

3.5.1.1 Improve the Quality of Clinical Evidence

The recommendations for the clinical trials for orphan advanced therapy medicinal products (ATMPs) were provided in the section "Strategies to Strengthen the Quality of Clinical Evidence for Orphan ATMPs." Additional suggestions for the optimized clinical evidence for ATMPs are on elaborated next.

TABLE 3.8

Price and Innovative Funding Models for RMs Marketed in the EU and USA

Products	Country	List Price	Payment Models	Details
Kymriah	UK	€320,000	Used within CDF and a managed access agreement	The data collection period is anticipated to conclude February 2023 when it is expected that the JULIET clinical trial is due to end to address the uncertainties in the overall survival, curative nature, use of IVIG, and duration.
	Italy		Outcomes-based, staged payments	1) manufacturer to repay treatment costs for non-responsive patients. 2) spreads payments, paying in three installments: upon enrollment in the treatment program, after 6 months of treatment, and after 12 months of treatment.
	Germany-GWQ	Outcome-based agreement		Novartis shares the risks of this arrangement by agreeing to partially reimburse these costs if the therapy result for survival is not achieved within a defined period of time.
	Spain	Outcomes-based, staged payments		One at the time of treatment (reported to be 52% of the total €320,000), and a second payment at 18 months (reportedly the remaining 48%), provided that the patient has achieved and sustained a complete response to the treatment.
	Japan	$305,800	Cost-based agreement	Although the peak sales for Kymriah are lower than the main threshold for requiring CEA following launch, it was selected for CEA provision due to the high price and low transparency of cost calculation. The results of CEA will be used for price reduction.

(Continued)

TABLE 3.8 *(Continued)*
Price and Innovative Funding Models for RMs Marketed in the EU and USA

Products	Country	List Price	Payment Models	Details
	USA	$475,000	Outcome-based agreement	-Novartis proposed an outcome-based contract with treatment centers, which is a voluntary agreement that only allows for payment when patients respond by the end of the first month. CMS pulled out of the plan for reimbursing either on an outcome-based payment or indication-based payment where Kymriah is priced at $373,000 for large B-cell lymphoma, while for pediatric leukemia, it is priced at $475,000.
Yescarta	Germany-VDEK	€327,000	Outcome-based agreement	Outcome-based discount agreement with VDEK that is initially valid for 2 years. The success of the therapy is measured by the duration of the patient's survival.
	UK		Use within CDF and managed access agreement	The data collection period is anticipated to conclude February 2022, when it is expected that sufficient data will have been collected to provide additional evidence to help resolve the uncertainties in the overall survival, progression-free survival, and IVIG use.
	Spain	Outcomes-based, staged payments		Payments linked to survival with a first payment of €118,000 and a second payment of €209,000.
	Italy		Outcomes-based, staged payments	Paid in three installments: the first payment is scheduled at 180 days after infusion, the second payment at 270 days, and the final payment at 365 days.
Luxturna	UK	£613,410	Commercial agreements	Simple discount patient access scheme

TABLE 3.8 *(Continued)*
Price and Innovative Funding Models for RMs Marketed in the EU and USA

Products	Country	List Price	Payment Models	Details
	USA	$850,000	1) Outcome-based payment: Harvard Pilgrim Healthcare 2) Innovative contract model: Express Scripts 3) Installment payment – ongoing discussion with CMS.	1) Outcome-based rebate aligned with long-term efficacy of Luxturna: Spark Therapeutics will share risk with certain health insurers by paying rebates if patient outcomes fail to meet a specified threshold, thereby linking the payment for Luxturna to both short-term efficacy (30–90 days) and longer-term durability (30 months) measures. The short-term and long-term measures will be based on FST testing scores. 2) Innovative contracting model: Spark Therapeutics would enter into an agreement with commercial payers under which the payer or payer's specialty pharmacy, rather than the treatment center, purchases Luxturna. As a part of this agreement, the payer agrees to provide coverage for its members consistent with FDA labeling of Luxturna, expedite benefits processing, and cap patient out-of-pocket amounts at in-network limits. 3) Installment payment that allows customers to pay for Luxturna in installments over several years rather than in a single, up-front payment.
Strimvelis	Italy	€594,000 $665,000	Pay for performances	GSK must return some portion of the funds paid by AIFA based on the terms of the agreement if it does not work. No further details found.

(Continued)

TABLE 3.8 *(Continued)*
Price and Innovative Funding Models for RMs Marketed in the EU and USA

Products	Country	List Price	Payment Models	Details
Zolgensma	Germany-GWQ		Pay for performances	Two relevant parameters, such as patient-relevant outcome parameters, have also been taken into account. In the event of a contract, AveXis assumes the risk of repaying in some cases up to 100% of the drug costs in phases.
	USA	$2.125 million	1) Installment payment 2) Outcome-based payment	AveXis has partnered with Accredo to offer insurers the possibility of paying in annual instalments of $425,000 for 5 years; AveXis is working closely with payers to create 5-year outcomes-based agreements.
Zynteglo	Europe	€1.575 million	Pay for performance with installment elements	The company proposed an installment plan, where the total cost will be spread over five years based on its continued effectiveness, at €315,000 a year.
Tecartus	UK	$373,000	Use within CDF and managed access agreement	More data from ZUMA-2 are expected, with additional years of follow up planned. The committee agreed that: • More data on progression-free, post-progression, and overall survival up to 5 years will help clarify if treatment with autologous anti–CD19-transduced CD3+ cells improves long-term survival. • Using autologous anti–CD19-transduced CD3+ cells in the NHS allows data to be collected using the Systemic Anti-Cancer Therapy (SACT) data set to get more accurate costs and benefits for its use in clinical practice, as well as the median age of the patients who would be offered treatment with autologous anti–CD19-transduced CD3+ cells.

- **Data gap for nature history data and limited clinical evidence**

For rare diseases, the paucity of data on disease characteristic has made it difficult for HTA bodies to appreciate the disease burden and course of illness and to determine the SOC and validity of surrogate endpoints, implying that epidemiology and natural history studies will be helpful to enhance knowledge and gain HTA acceptance (117). When feasible, manufacturers should explore the possibility of conducting at least one RCT comparing RMs with the best care available in which adaptive trial designs or weighted randomization could be considered in order to meet ethical requirements (61). Cluster randomization of centers may be also a good alternative to patient randomization. Patients may hesitate to participate in clinical trials because of the misconception about RMs or concerns about the disruption of daily activities. Therefore, patient education to destigmatize clinical trials of RMs could be a powerful tool to increase participation (110, 132). Moreover, to bridge the data gap, opinions from clinical experts could contribute to determining appropriate eligibility criteria in clinical trials and estimating long-term transition probabilities of diseases in economic analysis (71).

- **Surrogate endpoints**

Surrogate endpoints as a primary endpoint must be explicitly defined and justified. Drummond et al. suggested that the validity of surrogate outcomes could be examined from a three-level hierarchy: 1) biological plausibility, 2) the correlation between the surrogate and the final outcome across the individual or cohort level, and 3) evidence supporting the idea that the improvement in the surrogate outcomes is aligned with the final outcome in clinical trials (67). Systematic reviews to evaluate the association between the surrogate and the patient-relevant outcomes would be favored (89). Moreover, decision-makers may require explicit evidence showing improvements in surrogate outcomes are relevant to outcomes that are critical for reimbursement (e.g., QOL, disability, life expectancy, or QALYs) (119). Some HTA agencies do have defined clear recommendations or guidelines on the validation of surrogate endpoints, which are likely to be more demanding (133) but unrealistic to implement in several conditions compared with the aforementioned method. Therefore, to ensure acceptability of surrogate endpoints, manufacturers should consider full validation according to the most stringent HTA guidelines to ensure it will apply for all HTA agencies. If this turns out to be unfeasible, it should be documented, with reasons clearly explained, and alternative solutions such as those described herein could be implemented.

- **Indirect comparisons and extrapolations**

When randomization appears complex to implement, indirect comparisons could compensate for the limitations of a single-arm study (73) if the inherent bias in observational studies could be sufficiently adjusted. Berger et al. provided a detailed checklist to assess the credibility of observational studies (134). Indirect comparison will be more reliable when the population is homogeneous, confounding factors are well known, the primary endpoints are objective, the relative effect size is substantial, and the generalizability of historical data is acceptable (67). A prospective observational study with patients matching inclusion and exclusion criteria may be an alternative option that would be more acceptable than an adjusted historical comparison.

The best data source for indirect comparison might be a prospective patient registry with the study population matched with ongoing clinical trials (89). The patient registry should be 1) independent of manufacturers, 2) built on an internationally agreed upon set of data definitions to enable subsequent aggregation and meta-analysis, and 3) designed in an adaptive fashion to allow different countries to collect different, relevant evidence aligned with local requirements (76, 89) (Table 3.9). For example, the European Society for Blood and Marrow

TABLE 3.9
ISPOR Checklist in Evaluating a Network Meta-analysis

Section	Checklist Items
Introduction	Are the rationale for the study and the study objectives stated clearly?
Methods	Does the methods section include the following?
	• Description of eligibility criteria
	• Information sources
	• Search strategy
	• Study selection process
	• Data extraction (validity/quality assessment of individual studies)
	Is there a description of methods for analysis/synthesis of evidence? Do the methods described include the following?
	• Description of analysis methods/models
	• Handling of potential bias/inconsistency
	• Analysis framework
	Are sensitivity analyses presented?
Results	Do the results include a summary of the studies included in the network of evidence? Individual study data? Network of studies?
	Does the study describe an assessment of model fit? Are competing models being compared?
	Are the results of the evidence synthesis (ITC/MTC) presented clearly?
	Sensitivity/scenario analyses
Discussion	Does the discussion include the following?
	• Description/summary of main findings
	• Internal validity of analysis
	• External validity
	• Implications of results for the target audience

Source: Jansen J, et al. Interpreting Indirect Treatment Comparisons and Network Meta-Analysis for Health-Care Decision Making: Report of the ISPOR Task Force on Indirect Treatment Comparisons Good Research Practices: Part 1. License number for reuse: 4970140439733.

Transplantation has avoided siloed data being collected by insulated manufacturers or disease-focused groups, thus improving the efficiency and quality of evidence collection (95).

3.5.1.2 Think Economically Early in Development

- **Understanding the market demand and tailored approaches**

The specialty of RMs requires a transition from traditional technology-push commercialization pathways towards a market-pull model. In this model, technology developers must understand and target the evidence requirements of HTA agencies throughout the translation process, ensuring the existence of a "reimbursable evidence dossier" by the time of product launch (112, 135). Headroom analysis will be helpful for developers to understand the current market landscape for the targeted indications in terms of the competing products available, the reimbursement profiles of these products, and requirements for economic evidence (70). Without an in-depth understanding of these factors, RMs will have suboptimal patient access and commercial success (129). Moreover, to address the challenges in the fragmented markets with multiple players, tailored strategies targeting the specific evidence requirements of each payer segment will be meaningful (124).

Take the case of Imlygic (restricted reimbursement in England, but rejected in Germany): the selection of appropriate comparators should be decided after through consideration of the clinical practices and national guidelines of the targeted markets (74). In particular, RMs may span traditional lines of drugs and medical devices, whereas the HTA evidence requirements for medical devices and drugs could be markedly different. For example, when tisagenlecleucel was evaluated in Canada, it was reviewed by the body charged with evaluating medical devices (Health Technology Expert Review Panel) instead of the body that typically evaluates cancer therapies (Pan-Canadian Oncology Drug Review) (128). Finally, early engagement with payers and HTA agencies to understand how they define value is critical to incorporating the right endpoints into clinical and observational trials.

- **Explore methods for decreasing the cost**

Once the market landscape for the given indication is understood, this information can be used to define the economic criteria that must be satisfied in order to be successful, such as the upper limit of the manufacturing cost or the minimal effect size of clinical outcome (70). To minimize the manufacturing cost, the optimization of the manufacturing process and increase in scalability must be emphasized, such as the automated manufacturing procedure for allogenic RMs and the potential of universal CAR-T (136). Additionally, a study in Australia suggested that CAR-T manufactured in-house in academic centers is cost-saving without compromising the product quality (91). Yadav et al. indicated that hospitals working together with an academia center could be advantageous to reduce the time and cost of RM treatment (136) because of enhanced collaboration between experts from multiple scientific disciplines (e.g., oncologist, epidemiologist, and toxicologist) (119) and cost reduction through shared infrastructures (136). The possibility of RM administration in an outpatient setting necessitates further exploration; for example, it may be helpful in reducing the incidence of CRS related to CAR-T, thus reducing the cost of toxicity management (91).

- **Target the patients who could benefit the most from the treatment for value maximization**

Rather than assuming broad coverage, it is necessary to think whether there will be coverage restrictions depending on patient age, pre-existing comorbidities, or previous treatments (70). Therefore, identifying and targeting the patients with the highest likelihood of response will potentially increase drug value and HTA acceptance (91). Inclusion criteria of the clinical trials should be thoughtfully designed, with input from clinical experts and relevant stakeholders, in which a rigorous criteria review process using information collected from RWE and regulatory documents (e.g., FDA labeling) is recommended (67). For example, Bandeiras et al. suggested that stem cell therapy for type 1 diabetes would be most cost-effective in patients for whom end-stage renal disease is avoided. The development of predictive models of diabetes-related complications would assist RMs in better targeting these population (137). It will become increasingly important for companies to ensure the availability of companion diagnostic testing to identify patients with relevant gene mutations and who will benefit from a given therapy. Economic analysis combining RMs and companion diagnostics will be necessary to support reimbursement decision-making (138).

3.5.1.3   Implications for Decision-Markers

- **Landscape study**

Headroom analysis and horizon scanning could be helpful to deepen payer knowledge of the potential benefits in relation to the economic impact of emerging technology on healthcare systems (139), thus informing a better priority-setting strategy (123). In the face of substantial evidence uncertainties, value of information (VOI) analysis could be considered as an adjunct to HTA to determine whether future studies are necessary (76). VOI analysis uses Monte Carlo simulation to quantify uncertainty around the ICER and the risk of making the wrong decision. The expected value of further research will be a reduction in the expected cost of making the wrong decision (139).

- **HTA methodology considerations**

Several options can be used to explore how evidence uncertainties will impact the value assessment, such as the scenario analysis that was proposed in the US ICER value framework. Specifically, the US ICER Institute will develop two optimistic and conservative benefit scenarios to examine the uncertainties in the duration of benefit, magnitude of benefit, response rate, types of survival models, and relative benefit size (106). Introducing ethical and social issues of RMs into HTA should not be an arbitrary process, but a use of reliable sources to consider them systematically. This could include 1) seeking professional advice from bioethicists and social scientists, 2) conducting qualitative or quantitative primary research, and 3) performing secondary research integrating published literature (73). With the growing use of indirect comparison and RWE, additional effort is needed from the HTA bodies to standardize the methodology and advance their analytic capacity to make the best use of this indirect evidence in conjunction with formally elicited expert judgments (139).

- **Parallel dialogue with regulators and payers**

Regulators tend to adopt flexible approaches to evaluate the efficacy and safety evidence to assess the benefit-risk of applicants (internal validity), while payers focus on the relative effectiveness and safety in addition to cost-effectiveness to assess the additional health benefits over existing treatments (external validity) (58). Accordingly, regulators showed favorable attitudes towards accelerating the market approval of ATMPs despite immature evidence, in contrast to payers hesitant to reimburse ATMPs due to substantial uncertainties in long-term effectiveness, safety, and economic outcomes. Moreover, expedited approval programs may further widen the gap between regulators and payers considering the less stringent evidence for market authorizations were accepted (140). Such discrepancies in the evidence requirements of regulators and payers have partly contributed to the commercial failure of early-approved ATMPs in the EU, such as Glybera and Provenge. Both were withdrawn from the EU market because the "not-reimbursed" status threatened patient affordability and restricted market access.

Therefore, there is an urgent call for enhancing interactions and communications between regulators and payers in order to ensure that 1) all the evidence uncertainties (e.g., surrogate outcomes) with the innovative ATMPs have been sufficiently communicated (58); 2) agreements are reached regarding the drugs' qualifications for expedited approval programs based on the "unmet clinical needs" principle; and 3) the evidence requirements for pivotal trials supporting market approval, as well as qualities of confirmatory studies in the post-marketing obligations, will meet the expectations of both parties (141).

For example, in the EU, parallel consultations with EMA and EUnetHTA were initiated in July 2017, which allowed developers to obtain feedback on the evidence requirements for MA and reimbursement at the same time (142). This will enhance the developers to design the clinical trials and plan for a future commercial route at the early stage of development, thus reducing the possibility of raising major objections from either regulators or payers (143).

- **Alternative payment strategies**

The adoption of alternative payment arrangements (61) will necessitate intensive investment in the construction of a data infrastructure and sophisticated economic analysis to examine how breakthrough RMs will affect medical spending over time (127). The RAND Corporation suggests that two properties could be considered when deciding the optimal funding options for a given cure (127): the time to break even and the separability of the drug cost from the cost of overall disease management.

Payers and manufacturers have shared interests to work out more sustainable financing approaches to enable timely patient access to RMs (119, 144). Although the investments in infrastructure construction may outweigh returns in the short term, the long-term benefits could be substantial when more RMs targeting non-rare diseases come to market. For outcome-based payment, a collaborative attitude between payers and manufacturers is paramount to ensure relevant and realistic clinical milestones are defined and payment terms are mutually beneficial (97). To remove the legislative barriers in implementing the outcome-based payment, the US Center for Medicare & Medicaid Services (CMS) issued a proposed rule for "Best Price" in June 2020. Manufacturers, by law, must give Medicaid the "Best Price," which was historically defined as the lowest available price to any wholesaler, retailer, or provider. However, the "Best Price" rule can hamper the implementations of outcome-based payment considering that the drug would be free to provide to CMS (the best price is zero) in the case that the manufacturer offers a payer a money-back guarantee, but the drug turns out to be ineffective. Under the new rule proposed, more flexibility is allowed to report a multiple "Best Price" or a "Best Price" range if the drug is attached to outcome-based arrangements. Such initiative shows the CMS's attitude to encourage each state to enter into outcome-based payment for timely market access of innovative but expensive drugs (145).

In the case of installment payments, additional efforts are needed to address practical challenges, such as how to implement appropriate cross-payer transfers (146). Some legislative adaptations may be necessary to require all insurance companies to resume the remaining payment obligations of patients who switched plans via inclusion of payment obligations as part of a patient's "pre-existing conditions" (147). Drugs are consumable; their budget impact should appear on the budget of the year of administration independent of the payment schedule. Because RMs are often a one-time administration with long-lasting benefits, and drugs are considered a proxy for health production, RMs may qualify for amortization of an intangible asset if they can lead to economic values that flow from the asset in the long term. This will require adoption of a new standard at each jurisdiction level, as accounting standards are country specific. We are currently engaged in assessing the feasibility and the hurdles of using a revised version of the International Finance Reporting System standard combined with certain modifications in tax policy to integrate RM amortization.

Finally, it will become increasingly difficult to disconnect the cost-effectiveness analysis of RMs from the payment model because price may be affected by the product post-launch performance, and the discount rate should be aligned with the installment payment model that could be considered over a longer period of time (Table 3.10).

TABLE 3.10
Implications for Health Technology Assessment Agencies and Payers

Process	Challenges	Recommendations
HTA	Clinical evidence assessment	• Conduct scenario analyses (best-case scenario and worst-case scenario) to examine how the magnitude and durability of treatment benefits and survival model will impact the economic outcomes (57). • Standardize the methods and evidence requirements of defining and collecting post-market real-world evidence. • Coordination between countries for real-world monitoring to ensure greater consistency and efficiency in data collection. • Enhance the capacities of evaluating and incorporating indirect comparison in the HTA decision-making process (86). • Value of information analysis would be helpful to quantify the evidence uncertainties, inform the risk of making the wrong decisions, and determine whether additional evidence collection is necessary (86).
Economic analysis	Value assessment	• Make assessments of the social aspects of RMs in a robust way by resorting to experts from multiple disciplines (27). • Possible approaches include 1) seeking professional advice from bioethicists and social scientists, 2) conducting qualitative or quantitative primary research, and 3) performing secondary research integrating published literature (27). • MCDA can be used to obtain an overall estimation of additional value, thus complementing the CEA and make the HTA more robust and consistent, but more research to standardize the methodology to integrate the multiple value elements is required (8). • Provide two reference case analyses: a societal perspective for capturing broader benefits and a healthcare perspective which is more relevant for healthcare decision-making contexts (53).

(Continued)

TABLE 3.10 *(Continued)*
Implications for Health Technology Assessment Agencies and Payers

Process	Challenges	Recommendations
	Resource use	• Take additional costs related to the administration of RMs into account, such as the management of adverse events (14, 53).
	Model parameters	• More clarifications on the preferred methods for estimating indirect costs and how to incorporate them into decision-making. • Conduct sensitivity analyses that include scenarios with different discount rates (e.g., 0% or close to 0%) (8) and time horizons (e.g., lifetime), and consider the results of sensitivity analyses in their decision-making (53).
Payment	Landscape analysis	• Headroom analysis, horizon scanning, and cost-of-illness analysis could provide valuable insights on the potential value in relation to the possible economic impacts on the healthcare system of emerging technologies (73).
	Price	• Appropriate assessment on the price of comparators used for CEA; overpriced comparators will be more scrutinized (53). • Value-based payment will link more to budget impacts and "reasonable" profit margins (66).
	ICER	• Alternatives to a fixed ICER threshold include 1) a sliding scale for the ICER, 2) repricing cost offsets with clearly defined threshold, 3) QALY-cap equal to maximum WTP per QALY, and 4) shared saving between manufacturers and payers (8, 53, 66).
Alternative payment models		• Collaborative attitude between payers and manufacturers is paramount to ensure that relevant and realistic clinical targets are set for outcome-based payment and payment terms are mutually beneficial (18, 47). • The adoption of alternative payment arrangements will necessitate intensive investment in the construction of infrastructure and sophisticated economic analysis to examine how breakthrough RMs will affect medical spending over time (76).

TABLE 3.10 *(Continued)*
Implications for Health Technology Assessment Agencies and Payers

Process	Challenges	Recommendations
		• Two properties could be considered when deciding the optimal funding options for a given cure (76): the time to break even and the separability of the drug cost from the cost of overall disease management.
		• Modifications of regulations or legislation may be necessary to mitigate the challenges when patients switch plans (free-rider issues) and to avoid the adverse selection of patients (99).
		• For installment payments, legislative adaptations may require insurers to resume the payment obligations of patients who switched plans via inclusion of payment obligations as part of a patient's "pre-existing conditions" (99, 100).

CEA: Cost-effectiveness analysis; HTA: Health technology assessment; ICER: Incremental cost-effectiveness ratio; MCDA: Multiple-criteria decision analysis; QALY: Quality-adjusted life-year; RM: Regenerative medicine; RCT: Randomized controlled trial; WTP: Willingness to pay.

3.6 CONCLUSION

Manufacturers and payers have common goals to facilitate patient access to RMs that have substantial clinical and societal benefits. Improving the quality of clinical evidence will remain the key driver to reassure HTA agencies and increase chances of success. Having a clear perception of the evidence requirements for reimbursement of the targeted markets before embarking on the clinical trials will provide direction on where most efforts should be focused. Furthermore, manufacturers should explore ways to reduce costs of manufacturing and administration. From the HTA side, the necessity of implementing a specific HTA analysis framework for RMs is debatable. There are inconsistent opinions regarding the need for adjusting current principles for the reference case (e.g., differential discount rates) of economic analysis of RMs. However, it must be recognized that RMs pose a new problem for HTA agencies and payers by the concentration and the magnitude of well-known challenges that used to be seen more often in isolation. The increasing number of RMs to reach the market shortly with great potential to fulfill unmet medical needs but limited evidence and high prices will force HTA bodies and payers to reconsider their current decision-making process. Increased concertation of all stakeholders is unavoidable to delineate the future mutually beneficial strategies for RMs in order to ensure successful patient access, sustainability of insurance systems (public or private), and incentives for manufacturers to bring products needed by society. Amortization of RMs as a non-tangible asset may be a future option for addressing the high up-front cost of RMs, but more effort is necessary to study its feasibility and obstacles to implementation.

4 Impact of COVID-19 on the RMs Industry

Coronavirus disease 2019 (COVID-19), caused by a severe acute respiratory syndrome coronavirus 2 (SARS-CoV-2), has brought global damage to people's livelihoods and health since first reported in Wuhan, China, in late 2019 (148). As of 8 March 2021, 116,135,492 people had a confirmed infection and 2,581,976 people died from it all over the globe (149). The scale and severity of COVID-19 are unprecedented, which significantly disrupted social, economic, and political activities worldwide. The pharma industry, like other key economic sectors, has been severely disrupted. The disruption is particularly damaging for the cell and gene therapies (CGTs) industry due to its complexity in the manufacturing, supply chain, and clinical trials, in addition to the substantial challenges in price, reimbursement, and market access (5).

However, the COVID-19 pandemic also provided great opportunities for the pharma industry considering the public is desperate for a "miracle" drug and vaccines. CGT companies, with their extensive experiences in the research of cell biology, cellular immunity, genomic technology, and viral vectors, could have specific advantages to research and develop promising therapeutics for COVID-19. In this review, we first discuss how the COVID-19 pandemic has caused disruptions to the overall activities of CGT development, starting from manufacturing, all the way through to the health technology assessment (HTA) and reimbursement. In addition, the progress made during the pandemic in terms of potential CGTs for COVID-19–related disorders, financial investments that were raised for supporting the development activity, and CGTs newly launched for non-COVID–19 diseases are examined.

4.1 TREMENDOUS DISRUPTIONS FOR ALL ACTIVITIES

4.1.1 SUPPLY CHAIN AND MANUFACTURING

The supply chain of RMs, which was already logistically complicated before the pandemic (150), has had to face new challenges as the disease has rapidly evolved. The first challenge comes from the shortage of material supply for RMs (Figure 4.1).

That susceptibility is particularly true for cell-based therapy, which was manufactured either on an allogeneic or autologous basis. For allogeneic products, cell donors are less likely to come and donate due to physical distancing restrictions, and the US Food and Drug Administration (FDA) requires a 28-day investigation prior to donation for donors with a confirmed

DOI: 10.1201/9781003366676-4

Manufacturers
•The manufacturing and supply chain of RMs, which was already complicated before the pandemic, has had to face new challenges •Clinical trials for RMs are difficult to conduct because patients are unable to travel to the medical centers and the patient recruitment for rare diseases is more challenging •Staff shortages have delayed assay development and hence MA submission

Patients
•Patients are concerned about transmission of COVID-19: patients are unable to visit treatment centers due to travel restrictions •Patients are more worried about the toxicity related to CGTs than ever before •This is a time-consuming process for clinical staff, especially at the critical time, when there are shortages in personal protective equipment and hospital resources (beds and medical ventilators)

Regulators
•Struggling to keep up with their workload, even before the pandemic •Delays in the expected guidelines for RMs and ATMPs developers •The interaction with global regulators to create more harmonization of CGTs programs in different countries is challenging •CMC inspections can be affected by COVID-19 travel restriction

HTA bodies
•Small patient number: problems of missing data or censored data become more pronounced •Single-arm trials: become more evidence because alternative analytic methods were used •Placebo-controlled studies: lower willingness to participate •Indirect comparison: it is inappropriate to combine the clinical trials conduct during and after the pandemic •Outcome assessment: an accurate estimation of overall survival will be compromised in COVID-19 infected patients

FIGURE 4.1 Disruptions to the development of cell and gene therapies caused by COVID-19.

or suspected COVID-19 infection (151). For autologous products, cell collection from patients, the first step in the manufacturing process, requires human-to-human contact, including a visit to an apheresis center and interaction with apheresis technicians (152). However, many apheresis centers have suspended operations to limit the risk of exposure for the clinicians. Further, the manufacturing of autologous cell therapies (CTs) often relies on the shipment of fresh cell material from a centralized manufacturing facility within very tight timeframes (150). Disruptions to travel itineraries put those shipments at risk. For example, the ban on travel from Europe to the United States caused threats to the shipments of Novartis's Kymriah, the delivery of which is very time-sensitive (153). Although Novartis stated that it has found alternative methods to ship Kymriah, not all biotechnology companies are financially and technically competent to navigate such hurdles. About one-third of CT companies have reported delays or discontinuation of manufacturing activities in a virtual roundtable (154).

Regarding gene therapy (GT) manufacturers, their already constrained manufacturing capacity for viral vectors before the COVID-19 crisis may have become even tighter (155). This is due to the fact that many of the raw materials used to manufacture viral vector–based vaccines are the same as those used to manufacture GTs (156), thus raw materials were transferred to develop and manufacture what could be billions of doses of COVID-19 vaccines (157).

With the development of viral vector–based vaccines to combat COVID-19, Aledo-Serrano et al (158) raised a series of crucial questions regarding the following:

- What if the viral vectors become the mainstream strategy for vaccine development?
- Will the immune response elicited against the vector compromise the efficacy of future gene therapies?
- Should new treatment guidelines be implemented to facilitate orphan disease–affected patients to receive alternative types of vaccines?
- Should the youngest patients also be given the opportunity to be immunized with non-vector vaccines, since they are theoretically more susceptible to present gene-related diseases during their lifetime than older vaccine candidates?

These questions are paramount given that frequency of cross-reactivity among different AAV serotypes can be as high as 50%, while the presence of neutralizing antibodies against the viral vectors carrying a gene will preclude the patients being eligible for clinical trials.

4.1.2 CLINICAL TRIALS

Many non-COVID-19–related clinical studies have been halted to meet the requirements of the COVID-19 response or to reassign the medical personnel to combat the pandemic (159). It was estimated that at least 322 biopharma companies' trials have suffered as of July 2020. The majority of disrupted trials were phase 2 trials (44.8%), but early phase 1 (26.1%) and pivotal phase 3 trials (21.7%) were also affected (160). As FDA officer, Peter Marks, commented during one virtual meeting: "I'm worried about the phase 3 clinical trials or pivotal trials in gene therapies that are not going optimally" (161). Delays in pivotal studies imply that market approval of CGTs in the near term could be affected.

According to the GlobalData report, although the total number of trials disrupted by COVID-19 since June 2020 has declined slowly, the number of trials with difficulty in patient enrollment continues to increase (162). The slow enrollment could be more evident for clinical trials of CGTs. This is due to the fact that CGTs could only be administrated in a few specialized medical centers, sometimes requiring patients to travel long distances to participate (155). More than half of the CGT companies surveyed by the company McKinsey reported difficulties in recruiting patients or having to suspend trial enrollment to minimize the risk of contracting COVID-19 (154). In particular, COVID-19 raises specific operational challenges to clinical trials for CGTs targeting rare diseases, which already struggled to recruit enough patients before the crisis (110). A recent nationwide survey by the National Organization for Rare Diseases conducted with patients suffering from rare diseases showed that 39% of participants have faced challenges accessing medical care or treatment and 74% have had a medical appointment cancelled (163). Travel restrictions, patient concerns about being exposed to the novel coronavirus, and the withdrawal of non-essential services from healthcare centers have been contributing factors.

4.1.3 Delivering Therapies to Patients

The capacity to prepare and treat patients with CGTs in a hospital setting may be much more limited. CGT treatment sites must not only be audited and undergo Foundation for the Accreditation of Cellular Therapy accreditation but must also train personnel on procedures for sample collection, storage, and shipment (155). This is a time-consuming process for the clinical staff, especially at this critical time, when there are shortages in personal protective equipment and hospital resources are prioritized for COVID-19 patients.

The administration of CGTs also proved problematic. Hospitals and patients are concerned about the transmission of the novel coronavirus, with patients themselves unable to visit treatment centers because of travel restrictions. This is particularly true for CGTs considering that most target patients with more severe diseases and who are mainly immunocompromised. In addition, some CGTs, such as chimeric antigen receptor T cells (CAR-T), are associated with significant toxicities, such as severe cytokine release syndrome (CRS) and neurotoxicity (164). The potential toxicities usually need to be closely monitored and may require intensive care units, which have been overloaded with COVID-19–infected patients (165). Additionally, the current pandemic has exposed hospitals to supply shortages of tocilizumab, a monoclonal antibody for the management of CRS, because critical COVID-19 could also have the manifestation of CRS (152). This possibly leads to the delay in initiating new patients on CAR-T treatments. In addition, considering that inpatient administration of CGTs will be subject to diagnostic-related group (DRG) rules in some countries, these may not adequately cover all the treatment expenses of providing them on top of the budget allocated to COVID-19 (84). Indeed, financial constraints of hospitals and patients, aggravated by the COVID-19 crisis, may further limit some patients' access to the costly CGTs therapies (97).

4.1.4 Small Businesses

The disruption caused by the COVID-19 pandemic will be particularly challenging for small-scale biotech companies, who will struggle to finish or initiate their development programs (166). This may be further aggravated by reduced capital raised for continuing research efforts. That is because many start-up biotech companies are still at the pre-commercial stage, and therefore highly reliant on external funding. In the absence of product sales, decisions to invest will depend on pipeline progress, key data readouts, and specific research milestones (154). As a result of significant delays in the research, clinical trials, and data generation, some start-up biotech companies will struggle to raise more funding. It is uncertain how investors will respond to this, but likely they will make their decisions more wisely and focus on supporting only companies having CGTs with greater long-term value propositions (160).

4.1.5 REGULATORY ACTIVITIES

For CGT manufacturers, staff shortages have delayed assay development and hence submission of a biologics license application (BLA) to the FDA, which could in turn impact a CGT company's development and approval timelines (154). For example, despite success securing European Medicines Agency (EMA) approval of Zynteglo in the last year, Bluebird claimed in March 2020 that the completion of a rolling BLA submission to the FDA would not be anticipated until mid-2021 (167), and the BLA submission of LentiGlobin for sickle cell disease (bb1111) would be further pushed back to late 2022 because of new chemistry, manufacturing, and control requirements (CMC) of the FDA, as well as the optional delays of the partnering contract manufacturing organization (168).

Regulators are also struggling to keep up with their workload, even before the pandemic. Peter Marks estimated that he spent up to 75% of his time on GCTs a year ago, while now he spends about 80% of each day on COVID-related issues (e.g., vaccines and convalescent plasma) (166). The FDA acknowledges that the pandemic may lead to delays in the expected guidelines for regenerative medicine and advanced therapy developers in the areas of neurodegenerative diseases, genome editing, and CAR-T therapies, as well as in their efforts to streamline the development of the guidelines on "N of 1" therapies for ultra-rare disorders. Moreover, the interactions with global regulators trying to create more harmonization of CGT programs in different countries are also challenging (169). CMC inspections can be affected by COVID-19 travel restrictions as well, which made the FDA approval of Bristol-Myers Squibb's lisocabtagene maraleucel, as planned in November 2020, impossible (170).

The delays in regulatory response may have an unexcepted influence on the order of market launch, especially for the CGTs that are racing with multiple rival ones. For example, Sarepta has missed the goal of initiating phase 3 clinical trials for their gene therapy to treat Duchenne muscular dystrophy (DMD), SRP-9001–102, by the end of this year, because of a lengthier time waiting for FDA feedback than before (171). Such a delay could potentially enable Pfizer to lead the race to bring their DMD gene therapy, PF-06939926, to market earlier, thus broadening the approval gap between the two products. The economic damage of loss of opportunity to launch first could be profound for CGTs targeting rare diseases. The combination of small patient size and "curative" potential of CGTs may create a "winner-takes-all" dynamic, which means that once the initial "prevalence" of patients suffering from a particular rare condition is treated, the remaining market sales of followers will be limited to only newly diagnosed "incident" patients (172).

4.1.6 HEALTH TECHNOLOGY ASSESSMENT

HTA committees are also under-resourced, as many members, such as clinicians or allied health professionals, will not be able to participate in assessment meetings (173). The assessment of CGTs, due to unique biological characteristics and a scarcity of clinical evidence, will be highly demanding in terms of the opinions from very specialized clinical experts (71). Additionally, HTAs for non-COVID-19–related topics (including CGTs) will be delayed because faster appraisal for COVID-19–related topics is the first priority. For example, the National Institute for Health and Care and Excellence (NICE) has established the RAPID-C19 initiative, aiming to prioritize the treatment for COVID-19. This caused several months' pause of the HTA process for Zynteglo and Zolgensma (174, 175), while the timeline to finalize the appraisal process is still subject to uncertainty.

The complexity to make a reliable assessment for CGTs will be further amplified by the pandemic. This is because the clinical trials for CGTs are typically featured with small patient numbers (172), where the problems of missing data or censored data become even worse because of COVID-19. The inherent uncertainties in the single-arm trials of CGTs will become more evident because alternative analytic methods were used to adjust the missing data. Moreover, it will be more difficult to conduct comparative studies for CGTs targeting life-threatening, rare diseases, which mostly lack an alternative treatment available to serve as active comparator (71), while the patients' willingness to participate in placebo-controlled trials will be lower at this critical time due to the unpredictability of trial group assignment and an increased risk to virus exposure (110). It is also questionable whether an indirect comparison is appropriate to combine the clinical trials conducted during the pandemic with the clinical trials conducted before the pandemic (159). Moreover, the patients treated with CGTs, who usually suffer from severe conditions, will be at a higher risk of mortality and complications if they are infected with COVID-19 (176), thus an accurate estimation of the overall survival will be compromised. Another potential bias posed by the COVID-19 pandemic included incomplete follow-ups (possibly not at random) that may invalidate the planned analyses, and potential heterogeneity in the included patients may increase, especially for multicenter trials, as the prevalence/incidence data vary across regions, thus widening the confidence interval of outcome estimates. (177). All these factors will contribute to an increased uncertainty in the magnitude of treatment benefits of CGTs and the appropriateness of clinical inputs for the extrapolation model (71), making a robust HTA challenging.

4.2 PROGRESS MADE DURING THE PANDEMIC

4.2.1 NEW TREATMENT OPPORTUNITIES

Despite all the disruptions caused by COVID-19, it has made clear the importance of science and innovation to solve the puzzle. The biopharma industry plays a critical role as one of the few sectors that can really make a difference in the research and fight against COVID-19. This is reflected in a lot of companies, biotech or big pharma, that are working to come up with a treatment or a vaccine to counter COVID-19 (160). This pandemic has caused an unprecedented catastrophe for the world's populations and economies, but it seems to be having rather positive effects on the biopharma industry (178).

GT companies are particularly well-positioned to research and manufacture COVID-19 vaccines, as seen by the majority of the first vaccines for SARS-CoV-2 in clinical development being gene based, mostly messenger RNA (mRNA), DNA, and viral vector vaccines (179). The CRISPR/Cas system, which has evolved naturally in bacteria to defend against invading phages, also showed promises to be repurposed in mammalian cells to defend against RNA and DNA viruses (180), although a validation of its effect on SARS-CoV-2 in animal models is still needed.

In addition, CTs, such as mesenchymal stem cells (MSCs), could provide potential benefits for acute respiratory distress syndrome (ARDS) via mediation of immunomodulation and repair of damage to lung tissue (181). Based on our search of the American Society of Cell and Gene Therapy (ASGCT) clinical trials database, there were 94 clinical trials investigating CGTs for the treatment of COVID-19 underway up to March 2022. Most of them were in phase 1, phase 1/2, or phase 3 stages (Figure 4.2). Arabpour et al. (182) conducted a meta-analysis to study the safety and efficacy of stem cells in treating patients with COVID-19; the results suggested that stem cell therapy could significantly reduce the mortality rate (RR 0.471, 95% CI: 0.270–0.821) and morbidity (RR 0.788, 95% CI: 0.626–0.992) in patients with COVID-19 compared with the control group. The advancement in the cell therapy for the treatment of COVID-19 will enhance the knowledge on their mechanisms of action and safety profiles, which potentially will promote the development of CGTs for other diseases.

According to the annual report released by Cell and Gene Therapy Catapult in November 2020, the manufacturing space of the CGT industry in the United Kingdom (UK) has increased by 48% in 2020 compared to 2019, and several UK CGT developers are investigating viral vector–based vaccines to combat COVID-19 (183). Among them, Oxford BioMedical, which is a pioneer CGT company with a portfolio of LentiVector-based gene therapies and CAR-T therapies, has collaborated with AstraZeneca to advance their adenovirus COVID-19 vaccine, AZD1222, to phase 3 clinical trials. Another German CGT company, BioNTech, which is committed to developing mRNA products and CAR-T cell therapies for solid tumors, has collaborated with Pfizer to produce the mRNA vaccine, Comirnaty (BNT162b2), marking the approval of the first COVID-19 vaccine by the FDA (on 11 December 2020) and EMA (on 21 December 2020). Moderna, which is an mRNA company that has also engaged in gene editing therapy and regenerative medicines (184), has collaborated with the National Institute of Allergy and Infectious Diseases (NIAID), and the Biomedical Advanced Research and Development Authority (BARDA) to redirect its research efforts toward a COVID-19 vaccine, mRNA-1273, that was authorized by the FDA on 18 December 2020 (185). All these examples show the substantial underlying promises of CGTs, but

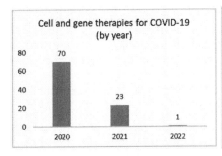
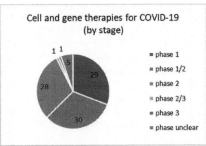

FIGURE 4.2   Cell and gene therapies for the treatment of COVID-19.

Note: Data collected using the clinical trials database of the American Society of Cell and Gene therapy; data as of March 2022.

also how CGT companies are responding quickly and shifting their strategies to explore more possibilities and create more value.

4.2.2 Financing Raised

Although start-up biotechs suffered dramatically in the pandemic, large biotechs with stable business lines do not seem to experience a shortage of capital. These biotech companies developing CGTs have had access to unprecedented levels of capital via venture capital (VC), initial public offering (IPO) routes. or equity investment, even during the period of the pandemic (169). As shown in the Alliance for Regenerative Medicine (ARM) report, in the first half of 2020, the regenerative medicine sector has already raised $10.7 billion, more than the total capital raised in 2019 and a 120% jump over the first half of 2019 (169). Speaking about COVID-19 from a strictly VC perspective, the financial chaos unleashed by the pandemic may have had far less of an impact on biotech than on other key sectors of the global economy. The likelihood of VC shortage for CGT clinical trials in the near term could be relatively low, especially for biotechs that have a sustainable pipeline (178). The reasons for this could be that investors in the biotechnology field are usually risk-tolerant and used to working with loss-making companies, and managing delays and hurdles is business as usual.

Using our internal database gathering partnership agreements (e.g., licensing agreement, manufacturing contract, joint venture) for CGTs, we investigated that the total up-front payment for partnership deals involving CGTs has already reached $1,024.90 million in the first half of 2020 (Figure 4.3), which considerably surpassed the up-front payment raised for previous years. This is still true in the second quarter of 2020, when the pandemic rapidly spread across the globe, indicating that the COVID-19 pandemic did not drive investors away from participating in CGT development-related activities. Additionally, merger and acquisition agreements related to CGTs made by big pharma continued to increase; the

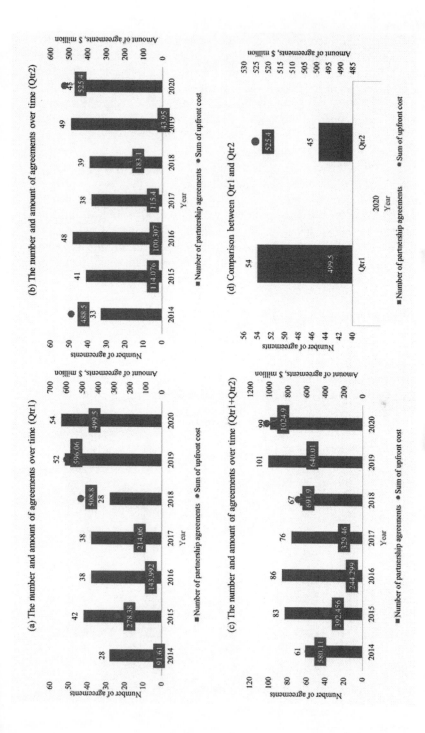

FIGURE 4.3 Partnership agreements for CGTs in the first half of 2020.

Qtr: Quarter.

notable ones included the acquisition of Allergan by AbbVie for $63 billion in May 2020, the acquisition of Asklepios BioPharmaceutical by Bayer for $4 billion in October 2020, and the acquisition of Prevail Therapeutics by Lilly for $1 billion in December 2020. In general, this figure for acquisition appears smaller than last year, predominately due to the acquisition of Celgene (by Bristol-Myers Squibb for $73 billion) and Shire (by Takeda for $62 billion). This is less surprising, since most CGTs are still in a relatively nascent stage of development (e.g., discovery or pre-clinical) and acquisition bears more risk than partnerships, which seems more obvious in this critical time.

In August 2021, the ARM, in partnership with GlobalData, released the report for the first half of 2021. ARM highlighted that 2021 was the "Year of firsts and records." According to the report, regenerative medicine and advanced therapies financing soared to new heights so far this year, raising $14.1 billion in the first half—already 71% of what was raised in all of 2020. The surge of investment in the sector has made H1 2021 the strongest half on record and put the sector on a path to outperform 2020, which broke financing records with nearly $20 billion raised despite the challenges of the COVID-19 pandemic (Figure 4.4).

4.2.3 REGULATION MILESTONES

4.2.3.1 Products Approved During the Pandemic

Despite all the distractions, there is still good news to hear that some CGT developers and regulators continue to manage advanced late-stage CGT candidates to the global market. For example, Tecartus, a CAR-T developed by Kite (a Gilead

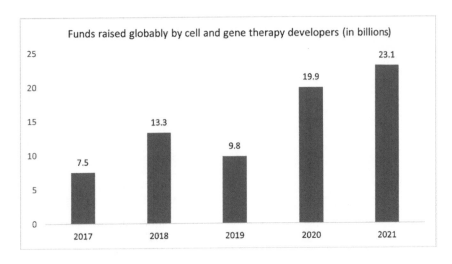

FIGURE 4.4 Total financing raised for regenerative medicine for the past 5 years.

Note: Data was collected as to the half year of 2021.

*Half year of 2021.

company) for large B-cell lymphomas, was approved by the FDA in July 2020 when the globe was severely hit by COVID-19. Year 2021 outperformed year 2020, with four gene therapies newly approved: Breyanzi, Abecma, Rethymic, and Stratagraft.

From the EMA side, Zolgensma, a gene therapy developed by AveXis (a Novartis company) for spinal muscular atrophy, was approved in May 2020. Tecartus and Libmeldy (developed by Orchard Therapeutics for metachromatic leukodystrophy) received positive recommendations for market authorization from the Committee for Medicinal Products for Human Use in October 2020. This was followed by the approval of Skysona and Abecma in July and August 2021, respectively.

Expedited programs strengthen the promising RMs to be launched in a faster manner (186), and the designation process was not hindered by the pandemic. From 2020 to 2021, 20 CGTs were granted with Regenerative Medicine Advanced Therapy (RMAT) designation by the FDA (Table 4.1), and 15 CGTs have been granted Priority Medicine (PRIME) designations (Table 4.2).

TABLE 4.1

CGTs That Received RMAT Designation in 2020–2021

Products	Manufacturers	Indication	Date
MultiStem	Athersys	Acute respiratory distress syndrome	23/09/2020
CTX001	CRISPR Therapeutics and Vertex	Severe hemoglobinopathies	11/05/2020
Ilixadencel	Immunicum AB	Metastatic renal cell carcinoma	06/05/2020
MDR-101	Medeor Therapeutics	Prevent kidney transplant rejection without chronic use of immunosuppressive drugs	22/09/2020
Kymriah	Novartis	Relapsed or refractory (r/r) follicular lymphoma	22/04/2020
Orca-T	Orca Bio	Blood cancers that are eligible for HSCT	15/10/2020
CD30-directed autologous CAR-T cell therapy	Tessa Therapeutics	Relapsed or refractory CD30-positive Hodgkin lymphoma	27/02/2020
TTAX02	TissueTech	Spina bifida in utero	16/04/2020
AB205	Angiocrine Bioscience	Organ vascular niche injuries	11/11/2020
Autologous Muscle Derived Cells	Cook Myosite	Stress urinary incontinence	17/12/2020
ALLO-715	Allogene Therapeutics	Relapsed/refractory multiple myeloma	21/04/2021

(*Continued*)

TABLE 4.1 *(Continued)*
CGTs That Received RMAT Designation in 2020–2021

Products	Manufacturers	Indication	Date
Valoctocogene roxaparvovec	BioMarin Pharmaceutical	Severe hemophilia A	08/03/2021
CTX110	CRISPR Therapeutics	CD19+ B-cell malignancies	21/11/2021
FT516	Fate Therapeutics	Relapsed/refractory diffuse large B-cell lymphoma	13/12/2021
ReNu	Organogenesis	Knee osteoarthritis	01/11/2021
RP-L201	Rocket Pharmaceuticals	Leukocyte adhesion deficiency-I	03/09/2021
Alofisel	Takeda	Perianal fistulas	02/10/2021

CAR: Chimeric antigen receptor; HSCT: hematopoietic stem cell transplant; PRIME: priority medicine; RMAT: regenerative medicine advanced therapy

[a] In 2020, a total of 13 RMAT designations were granted, only 10 of which that were publicly announced are listed in the table.

TABLE 4.2
CGTs That Received PRIME Designation in 2020–2021

Products	Manufacturers	Indication	Date
Adenovirus associated viral vector serotype 5 containing the *RPGR* gene	MeiraGTx	X-linked retinitis pigmentosa due to defects in retinitis pigmentosa GTPase regulator	27/02/2020
ADP-A2M4	Adaptimmune	Treatment of HLA-A *02–positive patients with inoperable or metastatic synovial sarcoma	23/07/2020
ALVR-105	AlloVir	Serious infections with BK virus, cytomegalovirus, human herpes virus-6, Epstein-Barr virus, and/or adenovirus in allogeneic HSCT recipients	30/01/2020
AT-GTX-501	Amicus Therapeutics	Pediatric patients with variant late infantile neuronal ceroid lipofuscinosis 6 (vlincl6)	17/09/2020
BB1111	Bluebird bio	Sickle cell disease	17/09/2020
CD30-directed genetically modified autologous T cells	Tessa Therapeutics	Classic Hodgkin lymphoma	17/09/2020

TABLE 4.2 *(Continued)*
CGTs That Received PRIME Designation in 2020–2021

Products	Manufacturers	Indication	Date
CTX001	CRISPR Therapeutics	Sickle cell disease	17/09/2020
OTL-203	Orchard Therapeutics	Mucopolysaccharidosis type I	17/09/2020
ARI-0001	N/A	Treatment of patients older than 25 years with relapsed/ refractory acute lymphoblastic leukemia	22/07/2021
ARU-1801	Aruvant	Treatment of sickle cell disease	25/03/2021
CT041	CARsgen Therapeutics	Treatment of patients with advanced gastric cancer who have failed at least two prior lines of systemic therapy	25/03/2021
CTX001	CRISPR Therapeutics	Treatment of transfusion-dependent β-thalassemia	22/07/2021
MB-107	Mustang Bio	Treatment of X-linked severe combined immunodeficiency (XSCID) in newly diagnosed infants	25/03/2021
AUTO1	Autolus therapeutics	Treatment of relapsed or refractory B-cell acute lymphoblastic leukemia	25/03/2021
RP-L201	Rocket pharma	Treatment of leukocyte adhesion deficiency-I	22/07/2021

4.2.3.2 Initiatives to Facilitate the Approval of Promising Products

Additionally, it could be encouraging to see that regulators have committed to facilitate the scientific review and approval timeline of treatment for COVID-19. The FDA has created a special emergency program, the Coronavirus Treatment Acceleration Program, with goals to use every available method to bring possible COVID-19 treatments to patients as quickly as possible, while at the same time validating whether they are helpful or harmful. The Emergency Use Authorization (EUA) is implemented when unapproved medical products or unapproved uses of approved medical products are allowed to be used in an emergency for life-threatening diseases or conditions when there are no adequate, approved, and available alternatives. As of 25 December 2020, the EUA has been granted to ten drug and biological products (including the BNT162b2 vaccine and mRNA-1273 vaccine, but excluding the revocation of hydroxychloroquine and chloroquine phosphate) (187). The EMA has established a COVID-19 pandemic task force to take quick and coordinated regulatory actions on the development, authorization, and safety monitoring of treatments and vaccines intended for the treatment and prevention of COVID-19.

Although no CGTs have been approved to treat COVID-19, companies that are developing CGTs for COVID-19 would benefit from these measures, including more intensive interactions with regulators to obtain scientific advice and facilitate clinical development. Considering the large number of COVID-19 patients to be treated and tested in clinical trials, it will provide more insights on the safety and effectiveness of CGTs in a faster way. On the other hand, albeit as one exceptional example, the delay of Zynteglo in the United States may provide new opportunities for Bluebird to seek approval for a broader patient population, including patients with β0/β0 genotypes and pediatric patients (168), rather than restricting it to patients older than 12 years old without a β0/β0 genotype, as allowed by the EMA. The implication is that COVD-19 may bring unexpected positive impacts to CGT manufacturers by offering breathing space for them to rethink their development plan and be better prepared for further disruptions.

4.3   CONCLUSION

The COVID-19 pandemic caused significant disruptions to the research, manufacturing, clinical development, and market launch of CGTs for non-COVID-19–related diseases. A shortage of material supply for manufacturing, difficulties in clinical trials (e.g., patient recruitment, conduct of placebo-controlled trials and follow-up), and a delay in regulatory dossier preparation all are contributing factors. This has emphasized the importance of addressing the challenges in the manufacturing and supply chain of CGTs to improve resilience in the crisis. Digitalization will improve the quality and accuracy of manufacturing and regulatory documents through increasing the traceability of the whole journey of CGTs. Telemedicine will be a powerful technique to mitigate the loss of follow-ups in clinical trials, especially for CGTs indicated for rare, life-threatening disorders.

Nevertheless, compared with other key global sectors, it is encouraging to see that some biopharma companies are shifting their focus and leveraging their decades of innovation experience to explore therapeutic options for the COVID-19 pandemic. Developers of COVID-19–targeted CGTs will benefit from more interactions with regulators to obtain timely scientific advice on their clinical programs, which will in turn support them to gain insights on how to proceed with their non-COVID-19 programs in the future. Developers of non-COVID-19–related CGTs are also advised to keep close relationships with regulators and HTA bodies to report the protocol deviation of clinical trials and better understand the possible adaptations in the evidence requirements.

5 Partnerships to Expedite Successful Market Access for RMs

As discussed in the previous chapters, RMs generally face unique and substantial challenges throughout their entire lifecycle (188), including the identification of suitable targets, manufacture of stable and high-quality products, collection of robust clinical evidence to demonstrate clinical promise, success in obtaining market authorization and reimbursement, and post-launch monitoring and tracing of patients being treated (189). It seems obvious that the successful market access of RMs could not be accomplished by one sole party, either small-sized biotechs or big pharma.

However, academia and start-up biotechnology companies, as the pioneers in the development of RMs (172), usually have limited financial capacity and experience to achieve the "bench to bedside" translation independently. Therefore, partnerships are critical for RMs to facilitate the development of discoveries and bring promising products to patients in need in a timely manner (172). Interests in partnership agreements are largely driven by the opportunity to merge complementary research and development (R&D) and business capabilities through which stakeholders can access potential therapies and innovative techniques, bolster pipelines rapidly, and raise capital for future business growth (190).

Owing to the potentially curative benefit of RMs for life-threatening diseases with limited (lacking) treatment options, it may be less surprising to see continued interest from biopharmaceutical companies to leverage partnership agreements to increase their footprint in this innovative and foreseeably lucrative market. Even the widespread disruptions and immense economic and health challenges unleashed by the COVID-19 pandemic seem to have had far less impact on biotechnology than on other key sectors of the global economy (178, 191).

Although it is well-recognized that the highest level of partnership activity in the RMs field has been in the last few years, studies analyzing the details of the partnership agreements are scarce. Questions remain unanswered regarding when the partnerships will take place, what the preferred forms of partnerships are, who the active participants of partnerships are, and how much the partnerships are worth. In order to understand how the partnership landscape of RMs has evolved over the past few years, we conducted an analysis of partnership agreements by comprehensively examining a range of factors related to the characteristics of products, including the product classifications, indications, and development

DOI: 10.1201/9781003366676-5

TABLE 5.1

Eight Types of Partnership Agreements

Collaborative R&D agreement	Long-term relationship in which each party contributes to R&D activities by utilizing their respective expertise.
Collaboration and licensing agreement	Partnership in which one party undertakes the development of products for another party in return for payment. One feature of this type of partnership is that one party is responsible for the early R&D activities until certain milestones (e.g., completion of phase 2 clinical trials) are reached. Afterwards, depending on which party holds the rights to the commercialization of candidates developed through the collaboration, the collaboration could be further classified as follows: • Commercialized by a single party, where sponsors assume the sole responsibilities of further regulatory and commercialization activities on a global basis • Co-commercialization whereby both parties share the cost and profits of subsequent activities, or each party has the right to commercialization in certain territories • Option licensing whereby sponsors acquire the (exclusive or non-exclusive) option to license the products but there is no certainty that sponsors will choose to exercise this option.
Licensing agreement	One through which one party grants its partner the access rights to an "object" in return for agreed-upon commercial terms, such as up-front payments, milestone payments, or royalty payments on the net sales of products. The licensed object could be a product, compound, technology platform, intellectual property, or a therapeutic target (e.g., cancer and genes). In this type of partnership, licensors will not participate in the subsequent R&D activities, and licensees will have sole responsibility.
Contract service agreement	One through which one party is contracted to provide services to another party. The services may include research, manufacturing, supply of materials, logistics, coordination of clinical trials, consultancy on regulatory or reimbursement evidence preparation, and personnel training programs.
Distribution and commercialization agreement	One through which one party is granted the rights to manufacture, market, distribute, and commercialize a product on behalf of another party in a given territory.
Joint venture agreement	One in which several parties set up a new company whose shareholding is held by the parties. Joint venture companies are generally intended to enable the joint development and subsequent commercialization of a technology or product.
Asset purchase	Acquisition and/or disposal of an asset owned by one party and desired by another party. An asset may include a product, trademark, technology, patent, manufacturing facility, or an entire business or division.
Grant	An award that one party gives to another party to facilitate the research, development, clinical trials, or other commercialization activities that are being conducted.

stage, and the characteristics of partnerships, including the size of the companies involved, partnership types, and financial terms. The typical types of partnership agreements are provided in Table 5.1.

5.1 OVERALL DESCRIPTION OF THE PARTNERSHIP FOR RMS

- **Year and amount of partnership**

From January 2014 to June 2020, 1,169 partnership agreements with a total value of $63,495.69 million (mean = $184.05 million) were identified. The year 2019 represents the highest number of agreements (N = 216) and the highest amount, with a total value of $15,785.25 million, of which the up-front value was $2,218.29 million (Figure 5.1).

- **Therapeutic areas**

The largest percentage of agreements concerning RMs was indicated for onco-logic diseases (N = 286; 25.31%), followed by neurologic (N = 123; 10.88%), opti-cal (N = 69; 6.11%), musculoskeletal (N = 55; 4.87%), and endocrine diseases (N = 52; 4.60%).

- **Stage of development**

At the moment of the partnerships, a majority of RMs (N = 739, 66.88%) remained in the early stage of discovery or pre-clinical development, 311 (28.14%) were being investigated in clinical trials, and 55 (4.98%) were under review by

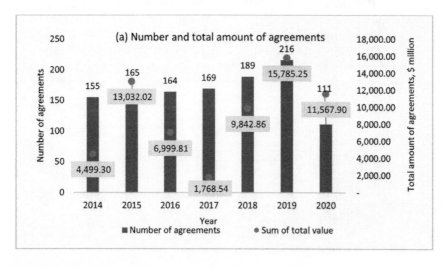

FIGURE 5.1   Number and amount of partnerships.

regulatory agencies or already approved in at least one country. The cumulative
number of agreements show that agreements concerning RMs in the early stage
of development have a faster growth rate than RMs in the clinical trial phase and
RMs in the regulatory or marketed phase.

- **Product classification**

Overall, gene therapy medicinal products (GTMPs) (N = 529; 45.25%) represented the largest number of partnership agreements compared with cell therapy
medicinal products (CTMPs) (N = 475; 40.63%). The number of agreements for
CTMPs was higher than those for GTMPs in 2014 and 2015, whereas the cumulative number of agreements for GTMPs exceeded those for CTMPs as of 2019.

- **Type of partnership agreements**

The most common type of partnership was a collaborative R&D agreement
(N = 339; 29.00%), followed by licensing agreements (N = 200; 17.11%), collaboration and licensing agreements (N = 199; 17.02%), contract service agreements
(N = 141; 12.06%), and grants (N = 141; 12.06%). Regarding the right to the
commercialization of the discoveries developed under collaboration and licensing
agreements (N = 199), most will be licensed to and commercialized by a single
party (N = 97; 48.74%) (Figure 5.2).

Regarding the goals of the partnership agreements, the majority aimed to
jointly develop RMs for the treatment of diseases (N = 549; 46.96%), to develop
cell engineering or processing technology (N = 184; 15.74%), to license manufacturing facilitates (N = 71; 6.07%), and to develop gene delivery vectors (viral
or non-viral) (N = 67; 5.73%) for GTMPs.

In regard to the relationship between classification of products and type of
partnership, the most common type of partnership for CTMPs (N = 475) was a
collaborative R&D agreement (N = 166; 34.95%), while for GTMPs (N = 529),
it was a collaboration and licensing agreement (N = 155; 29.30%). CTMPs have
a higher percentage of collaborative R&D, grants, contract services, distribution
and commercialization, and asset purchase agreements than do GTMPs, while

FIGURE 5.2   Types of partnership agreements.

GTMPs have a higher percentage of collaboration and licensing agreements, licensing agreements, and joint venture agreements than do CTMPs.

Regarding the relationship between development stage and type of partnership, for RMs in the early stage of discovery and pre-clinical development (N = 739), most partnership agreements are collaborative R&D agreements (N = 259; 35.05%), followed by licensing agreements (N = 169; 22.87%), and collaboration and licensing agreements (N = 159; 21.52%). For RMs being investigated in clinical trials (N = 311), most of the partnership agreements are contract service agreements (N = 74; 23.79%), followed by collaborative R&D agreements (N = 72; 23.15%) and grants (N = 61; 19.61%). For RMs that were under review by regulators or already marketed (N = 55), most of the partnership are distribution and commercialization agreements (N = 29; 52.73%), followed by contract service agreements (N = 9; 16.36%) and asset purchase agreements (N = 5; 9.09%) (Figure 5.3).

Regarding the relationship between type of partnership and amount of agreements, collaboration and licensing agreements (total = $47,822.46 million, mean = $487.98 million) contributed to 75.32% of the total amount of partnership agreements (total = $63,495.69 million, mean = $184.05 million) (Figure 5.4).

The total up-front amount of the partnership agreements was $7,158.21 million (mean = $29.34 million) of which 51.59% was contributed by collaboration and licensing agreements ($3,693.14 million, mean = $170.00 million). Cumulative analyses of the four most common types of partnerships showed that collaboration and licensing agreements consistently have the highest amount (total value and up-front value) and the fastest growth (in amount) compared with licensing agreements, grants, and contract service agreements (Figure 5.5).

- **Structure of participants**

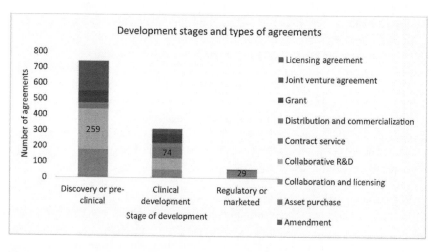

FIGURE 5.3 Development stage and types of agreements.

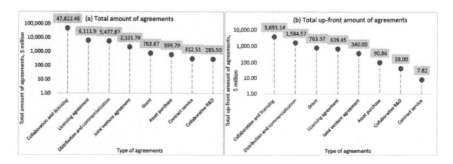

FIGURE 5.4   Type and amount of partnerships.

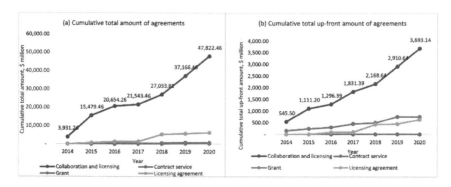

FIGURE 5.5   Amount and types of agreements.

Overall, the majority of partnerships occurred between biotechnology and not-for-profit institutions (N = 416; 35.59%), followed by biotechnology and biotechnology institutions (N = 209; 17.88%), biotechnology and service companies (N = 193; 16.51%), and biotechnology companies and the major pharmaceutical companies (N = 134; 11.46%). For the partnerships between biotechnology and not-for-profit institutions, academia (N = 295; 70.91%) was more involved than governmental organizations and charitable organizations. The cumulative number showed that partnerships between biotechnology and not-for-profit institutions have the fastest growth than other partnerships involving biotechnology companies. Among these, partnerships between biotechnology companies and academia have the fastest growth versus biotechnology companies and government agencies or biotechnology companies and other charitable organizations (Figure 5.6).

Regarding the relationship between the structure of participants and the type of partnership, partnerships between biotechnology companies and biotechnology/not-for-profit organizations/service companies (led by biotechnology companies) and partnerships between biotechnology companies and the major pharmaceutical companies (led by the major pharmaceutical companies) were compared.

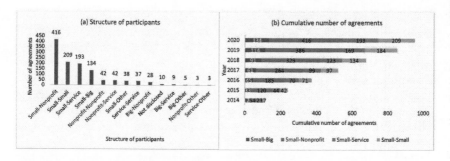

FIGURE 5.6 Structure of partnership agreements.

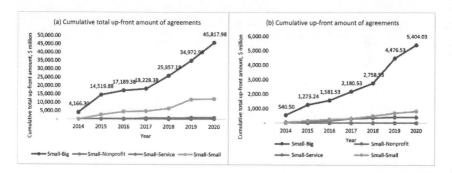

FIGURE 5.7 Type and amount of partnerships.

Collaborative R&D agreements accounted for the largest percentage (32.23%) of agreements led by biotechnology companies. Collaboration and licensing agreements accounted for the largest percentage (46.11%) of agreements led by the major pharmaceutical companies.

Regarding the relationship between structure of participants and amount of agreements, cumulative analyses showed that agreements led by major pharmaceutical companies consistently have a higher amount (total cost and up-front cost) and faster growth (in amount) compared with agreements led by biotechnology companies (Figure 5.7).

5.2 KEY PLAYERS IN PARTNERSHIP AGREEMENTS

- **Companies with Ten or More Partnership Agreements**

Nineteen companies in the pharmaceutical industry have established ten or more partnership agreements: 14 biotechnology companies, 2 major pharmaceutical companies (Takeda and Pfizer), 2 biotechnology companies (Kite Pharma and Spark Therapeutics) that were acquired by major pharmaceutical companies, and 1 contract manufacturing company (Lonza).

Five of nineteen companies have RMs already approved in at least one country: Kite Pharma (Yescarta [axicabtagene ciloleucel] and Tecartus [brexucabtagene autoleucel]), Bluebird Bio (Zynteglo), Takeda (Alofisel [darvadstrocel]), Spark Therapeutics (Luxturna [voretigene neparvovec]), and Orchard Therapeutics (Strimvelis). Twelve of nineteen companies have RMs currently being investigated in clinical trials, among which Bluebird Bio (N = 8), Sarepta Therapeutics (N = 6), Orchard Therapeutics (N = 6), and Caladrius Biosciences (N = 5) have more than five RMs in clinical development in their pipelines (Figure 5.8).

Among these 19 companies, Kite Pharma paid the highest amount ($4,332.50 million) to its partners, followed by Pfizer ($4,284.80 million), Bluebird Bio ($1,299.50 million), and Takeda ($1,035.25 million). CRISPR Therapeutics ($2,925.00 million) received the highest amount from its partners, followed by Sarepta Therapeutics ($2,850.00 million), ViaCyte ($2,065.50 million), and Kite Pharma ($1,360.00 million).

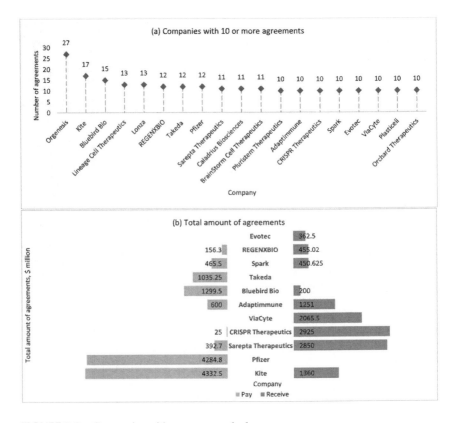

FIGURE 5.8 Companies with ten or more deals

Among not-for-profit organizations, the California Institute for Regenerative Medicine (CIRM; N = 26) had the largest number of agreements, followed by the University of Pennsylvania (N = 20) and Cell and Gene Therapy Catapult (N = 15). Among the top ten not-for-profit organizations with the highest number of agreements, CIRM ($285.39 million) paid the highest amount for grants, followed by the US National Institutes of Health ($214.73 million) and the Israeli National Authority for Technological Innovation ($20.81 million). The University of Pennsylvania ($2,119.30 million) received the highest amount from its sponsors, followed by Cell and Gene Therapy Catapult ($83.98 million) and the University of Massachusetts ($75.30 million).

- **Agreements with Amounts Larger than $100 Million**

A total of 90 agreements with a total amount larger than $100 million were investigated, of which 75 (83.33%) targeted GTMPs and 12 (13.33%) targeted CTMPs. The possible rationales behind the high amount of agreements were summarized into technology factors (e.g., chimeric antigen receptor [CAR] off-the-shelf products, and novel gene vectors) and non-technology factors (e.g., large market and RMs approaching market approval).

Regarding the type of partnerships agreements, collaboration and licensing agreements accounted for the largest number (N = 60; 66.67%) and highest total amount ($46,516.55 million).

Regarding the structure of the participants, partnerships between biotechnology and the major pharmaceutical companies accounted for the largest number (N = 55; 61.11%) and highest total amount ($44,731.30 million).

Companies that were most involved in these 90 agreements were further analyzed. Biogen paid the highest amount ($6,123.00 million) to its partners, followed by Vertex ($4,785.00 million) and Johnson & Johnson ($4,520.50 million). Sangamo received the highest amount of payment ($6,338.30 million) from its partners, followed by Cellectis ($4,123.20 million) and Voyager Therapeutics ($3,194.00 million).

Not-for-profit organizations were involved in only two agreements with an amount higher than $100 million: the University of Pennsylvania received $2 billion from Biogen to advance the development of gene therapy and gene-editing technologies that target the eye, skeletal muscle, and central nervous system (CNS) diseases; The US National Institutes of Health and the Bill & Melinda Gates Foundation invested $100 million each to develop affordable, gene-based cures for sickle cell disease and human immunodeficiency virus.

- **Engagement of the Major Pharmaceutical Companies**

Among the top 50 major pharmaceutical companies investigated, 28 of have established at least one partnership agreement. Takeda (N =14) has the largest number of partnership agreements, followed by Pfizer (N = 13) and Merck

(N = 11). Biogen ($6,128.00 million) paid the highest amount to its partners, followed by Pfizer ($5,124.80 million) and Roche ($4,930.00 million) (Figure 5.9).

A total of 26 biotechnology companies that specialize in cell and gene therapies have been acquired by the major pharmaceutical companies. Six of these

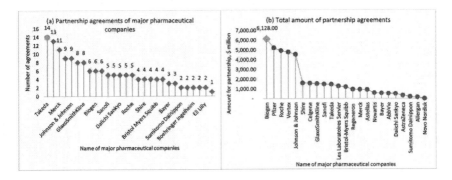

FIGURE 5.9   Partnership agreements of major pharmaceutical companies.

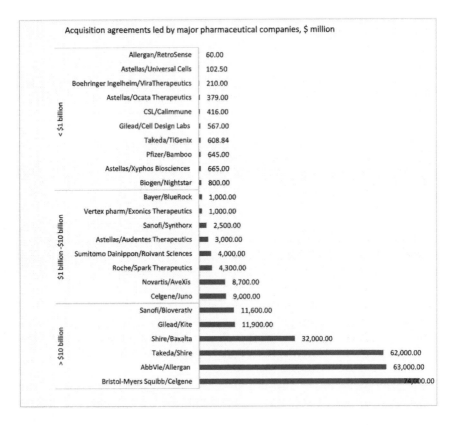

FIGURE 5.10   Acquisition agreements of major pharmaceutical companies.

twenty-six acquisitions have an amount higher than $10 billion; the highest payment was for the acquisition of Celgene by Bristol-Myers Squibb for $74 billion. Eight acquisitions have an amount of $1 billion to $10 billion; the highest payment was for the acquisition of Juno Therapeutics by Celgene for $9 billion. Nine acquisitions have an amount between $100 million and $1 billion; the highest amount was for the acquisition of Nightstar Therapeutics by Biogen for $800 million. One acquisition has an amount lower than $100 million, and two acquisitions did not disclose the financial terms (Figure 5.10).

Considering the total payment of combining partnership agreements with acquisition agreements, seven companies placed emphasis on both partnership and acquisition: Bristol-Myers Squibb, Takeda, Shire, Sanofi, Celegene, Roche,

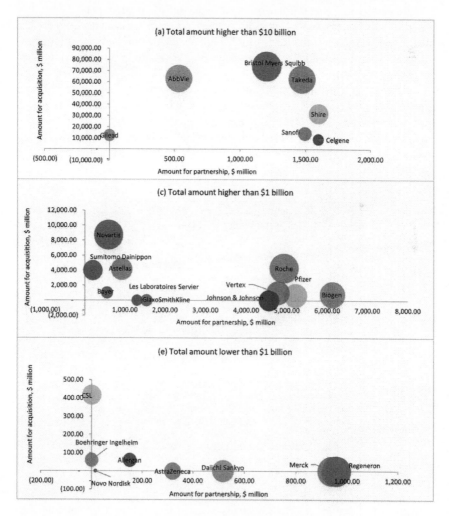

FIGURE 5.11 Total amount of partnership and acquisition agreements.

and Vertex. Five companies placed more emphasis on partnerships: GlaxoSmith-Kline, Johnson & Johnson, Pfizer, Biogen, and Les Laboratoires Servier. Six companies placed more emphasis on acquisitions: AbbVie, Gilead Sciences, Novartis, Astellas Pharma, Bayer, and Sumitomo Dainippon (Figure 5.11).

5.3 DISCUSSION

5.3.1 Overall Description

The number of agreements grew at a rate of approximately 10% annually, reached a spike of 18.5% growth in 2019, and it is still on track to maintain that same level of growth in 2020.

Partnership agreements concerning early-stage RMs predominate, a trend that has continued since 2014. This trend reflects not only the substantial technical challenges presented in the R&D of RMs but also the extensive interest of investors and other organizations, even though most RMs are in a relatively nascent stage (Figure 5.12).

Oncology remains the top therapeutic area for partnerships; however, more efforts have now been focused on solid tumors, which are typically more prevalent than blood tumors. Oncology drugs were offered with an opportunity for high prices, eligibility for expedited approval programs, and supportive initiatives such as National Cancer Plans and the Cancer Drugs Fund in the United

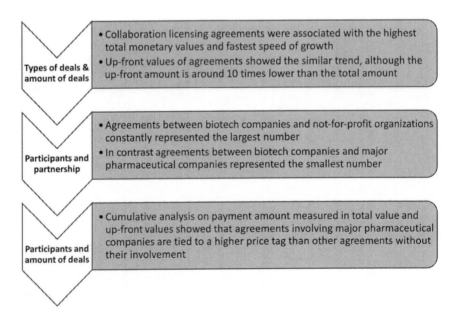

FIGURE 5.12 Overall descriptions of the partnership agreements for cell and gene therapies.

Kingdom (UK) (192). Neurologic disorders are the second most attractive therapeutic area for partnerships. Indeed, CNS disorders may lead to systematic dysfunction and severe disabilities. However, most drugs for CNS disorders have failed to provide relevant benefits to patients for many reasons, such as the complexity of the CNS, the limited regenerative capacity of the tissue, and the difficulty in conveying conventional drugs to breach the blood-brain barrier (193). It is estimated that 89.6% of pediatric rare genetic disorders compromise the CNS (194), which makes the CNS a likely target for gene therapies. Cell therapies have also attracted attention in neurodegenerative diseases (e.g., Parkinson disease) because of their potential for reinnervation of the neuronal network and neuro-restoration, allowing disease modification instead of palliative outcomes (195).

CTMPs initially dominated the market, primarily mesenchymal stem cells (MSCs), while recent years have seen a flurry of GTMPs, such as CAR-T cell therapies and gene replacement treatment. These developments have resulted from improved knowledge of the underling genetic causes of many rare diseases and the dissemination of knowledge about GTMPs from academic and other public research institutions to start-up biotechnology companies (188).

The cumulative total amount of partnership agreements represented $63.5 billion, an overall 14-fold increase from 2014 of $4.5 billion. The cumulative up-front amount represented $7.2 billion, an overall 10-fold increase from 2014 of $0.71 billion but only 11.3% of the cumulative total amount. This relatively low up-front amount also illustrates the early R&D stage of RMs targeted by these agreements. Collaboration and licensing agreements provided the highest total amount and up-front amount compared with other types of partnerships.

The major pharmaceutical companies engaged in more licensing agreements than pure collaboration agreements, while biotechnology companies more frequently engaged in collaboration agreements without licensing agreements. Agreements involving versus not involving the major pharmaceutical companies are tied to a far higher price (by three- to four-fold), indicating that these companies tend to take a more proactive approach to forging these partnerships.

5.3.2 Types of Partnership Agreements

- **Collaboration agreements**

This study shows that collaborative R&D agreements (without licensing elements) predominated the licensing agreements (with or without collaboration). This could reflect the fact that the majority of RMs were still in the early discovery period or in pre-clinical trials, and thus attracted more interest from small companies with the capabilities of advancing products to the next developmental milestone without requiring funding. Only later would licensing agreements be considered and involve larger players (e.g., the major pharmaceutical companies) with higher payment values.

Providing an overview of the landscape of acquisition agreements is not the primary scope of this study. However, previous studies suggest that collaboration agreements for RMs have been preferred by the biopharmaceutical industry over outright mergers or acquisitions, which are frequently executed for other types of innovative therapies such as monoclonal antibodies (mAbs) (196). Such conservative attitudes towards RMs were also reflected in the "adaptability" of collaboration agreements, where the option to license or commercialize the products would only be exercised when pre-agreed outcomes are delivered. This shift may be partly due to the substantial uncertainties about long-term safety and efficacy data associated with RMs and the unpredictability of future regulatory and reimbursement decisions. Thus, collaboration can be a lower-risk choice for investors who want to expand their RM portfolios without assuming a huge financial responsibility.

- **Partnership Agreements in Manufacturing and Logistics Management**

Manufacturing bottlenecks are still one of the main challenges in clinical trials and the commercial use of RMs (197). Confirmation of an optimal delivery platform for RMs will require specialized clinical expertise and advanced manufacturing capabilities that are not typically found outside of major biotechnology hubs. For example, only a small number of biotechnology and major pharmaceutical companies have adequate expertise and capability to produce optimized viral vectors for gene delivery on-site. This study shows that the mission of approximately half of the agreements was to advance or obtain access to technology platforms (e.g., gene editing, cell processing tools), either through collaboration agreements, licensing agreements, or manufacturing contract service agreements with third parties.

Apart from the collaborations for upstream R&D activities, partnerships for downstream activities related to the delivery of RMs to patients were also forged. The logistic process of RMs, ranging from initially sourcing raw materials from patients to tracking products until they safely reach the patient, can be extremely resource-intensive and require specialized technological support (e.g., cryopreservation) (Figure 5.13). Therefore, it is reasonable to see a number of contract services aim to support the materials sourcing, tracking, and delivery of RMs. Additionally, collaborations with digital or software companies are also powerful for tracing and recording the entire end-to-end patient journey in an efficient and automatic manner. Considering the challenges in clinical trials for rare genetic diseases, the biopharmaceutical industry has also partnered with Contract Research Organizations (CROs) to streamline the entire clinical development process by taking advantage of their medical expertise, strong experience in clinical trials, and data management and statistical capabilities. Albeit uncommon, partnerships are reaching even further into post-marketing activities, such as reimbursement. For example, Bluebird Bio has participated in a consortium led by Duke University, Center for Health Policy to develop a value-based payment model for RMs and other innovative treatments.

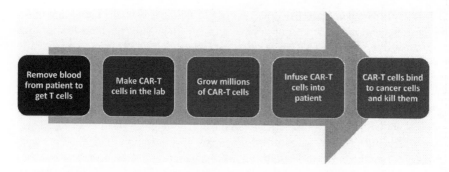

FIGURE 5.13   A typical manufacturing process of CAR-T cell therapies.

Source: US National Cancer Institute: www.cancer.gov/publications/dictionaries/cancer-terms/def/car-t-cell-therapy

- **4.3 Active Markers of Partnership Agreements**
  - Not-for-Profit Organizations

One unique feature related to the development of RMs is that not-for-profit institutions (e.g., academia and governmental agencies) are more extensively involved (198). Kassir et al. suggested that 60% of clinical trials for gene therapies in the United States were sponsored by not-for profit institutions (10% by the National Institutes of Health, 25% by universities, and 25% by hospitals) (199).

From our observations, all approved RMs in the United States and the European Union were initially discovered and developed in academia and were licensed to pharmaceutical companies depending on the timing of certain milestones. However, the drug development model that was prominent in the past, where academia performed basic science and discovered suitable targets and then pharmaceutical companies licensed and developed drug candidates, differs from the drug development model for RMs. First, technology transfer from academia to the pharmaceutical industry is more complex for RMs because agreements must ensure that donor informed consent is not violated at any stage of downstream product administration (200). Second, rather than simply being involved in the discovery and pre-clinical stage, academia and the pharmaceutical industry are collaborating more often in the downstream clinical applications in terms of possible protocol modifications at the time of administration (136). Third, and more importantly, active academic participators have advanced their efforts further by establishing their own spin-off companies to lead the development and commercialization independently. Two examples are Tigenix as a spin-off company of the Katholieke Universiteit Leuven and the Universiteit Gent for the marketing of Alofisel (darvadstrocel) and Spark Therapeutics as a spin-off company of the Children's Hospital of Philadelphia for the marketing of Luxturna (voretigene neparvec).

Another fundamental difference in the RM field has been the increasing interest from government institutions (198). As the most generous funders have shown in this study, the CIRM has made significant investments in RMs in both academia and biotechnology companies, spanning the discovery, translation, research, and clinical trials stages (201). The European Union has offered attractive funding opportunities for innovative technologies, including RMs, through its Innovative Medicines Initiative and its Horizon 2020 €10 billion funding program (198). Innovate UK, funded by the UK government, has supported Cell and Gene Therapy Catapult (202) as a center of excellence in innovation, with the core purpose of building a world-leading cell and gene therapy sector in the UK as a key part of a global industry. Government funding plays a critical role in supporting early-stage research and can help scientific advances progress to a stage where the level of risk is acceptable to private players. A second reason for the substantial interest from government is its ambition to take the global leadership position in this promising field. With RMs offering enormous commercial potential, there are strong incentives for government to help generate the infrastructure needed in support of science (203).

• Biotechnology Companies

The business model for a start-up biotechnology company could be a technology-driven-product model. This model involves first identifying new potential applications of the technology/discovery platform, either through internal efforts or direct licensing from academia. Afterwards, a start-up biotechnology company signs collaboration and licensing agreements with strategic alliances to advance and commercialize the proprietary products in exchange for investments from collaborators (204). As a result, the biotechnology company can enjoy stable revenues from licensing or sales, which allows for attracting more investors or utilizing such an income stream to develop its wholly owned products. This strategy creates a "win-win" situation where investors benefit from reduced risks by taking only part, not full, responsibility, and the start-up biotechnology company benefits from near-term revenue generation to maintain business growth (204).

This business model is feasible because many RMs hold promising opportunities in the diversification of one platform that are not possible with small molecules or mAbs engineered to target a single pathway (196). One successful example is REGENXBIO, which by utilizing its proprietary novel AAV vectors (NAV) technology platform for gene delivery has established multiple partnerships with both biotechnology and the major pharmaceutical companies. These partnerships have enabled REGENXIO to receive a milestone payment from AveXis (now Novartis Gene Therapies) upon the approval of Zolgensma (onasemnogene abeparvovec-xioi) in the United States. In turn, REGENXIO is developing internal gene therapy candidates targeting retinal, metabolic, and neurodegenerative therapeutic areas. Another example is Sangamo Therapeutics. With its zinc finger nuclease (ZFN) gene-editing platform, Sangamo Therapeutics has attracted significant investments from multiple major pharmaceutical companies, contributing to 18 product candidates (in partnership or wholly owned) in development. The opportunity to apply a platform to different therapeutic areas can help mitigate risk and expand the commercial potential of that platform.

- Major Pharmaceutical Companies

Historically, because the major pharmaceutical companies have been risk-averse and profit-driven, they were hesitant to engage when the RMs field was emerging due to skepticism surrounding the clinical benefits and commercial prospects of RMs. Nevertheless, the engagement of the major pharmaceutical companies in RMs has tended to come faster and occur earlier compared with other innovative therapies. For example, although mAbs took more than 20 years to attract extensive attention from the major pharmaceutical companies (196, 205), their interest in the RM field had begun to increase after 2010, just as proof-of-concept was validated (206). This study shows that the major pharmaceutical companies have never been key players when measured in number of agreements; however, they are the biggest sponsors who are more oriented toward licensing agreements with high value amounts. It was estimated that 16 of the world's largest (by revenue) 20 major pharmaceutical companies now have gene or cell therapy assets in their product portfolios (207).

Once the major pharmaceutical companies decided to invest in the field of RMs, they had three main options: build a proprietary platform in-house, buy an existing platform, or form a partnership to gain access to platforms developed by others (207) (Figure 5.14). Because partnerships in the still-nascent RMs sector are relatively inexpensive, the major pharmaceutical companies can afford to hedge their bets by spreading them over several partnerships where future success might lie. There has been a clear preference for the major pharmaceutical companies to enter into partnering agreements to explore RMs initially rather than pursuing the straightforward acquisition route or originating assets in-house (206, 207).

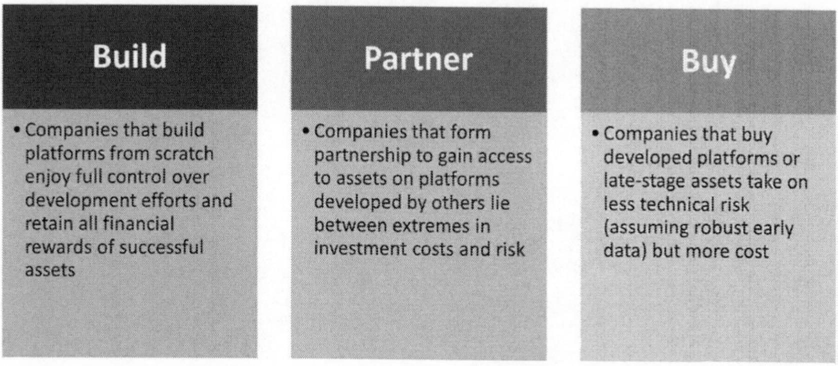

Build	Partner	Buy
• Companies that build platforms from scratch enjoy full control over development efforts and retain all financial rewards of successful assets	• Companies that form partnership to gain access to assets on platforms developed by others lie between extremes in investment costs and risk	• Companies that buy developed platforms or late-stage assets take on less technical risk (assuming robust early data) but more cost

FIGURE 5.14 Option for investment in the RMs field.

* This figure was adapted from McKinsey Company: Biopharma portfolio strategy in the era of cell and gene therapy.

As the technology underlying RMs matures, especially after the launch of several RMs with high price tags, the major pharmaceutical companies become more enthused about owning the technology rather than partnering (208). This fact is reflected in the buyouts with multibillion-dollar values, mostly targeting biotechnology companies with late-stage RM assets with a more favorable benefit-to-risk profile. For example, Gilead Sciences acquired Kite Pharma (Yescarta), Novartis acquired AveXis for Zolgensma, and Celgene acquired Juno Therapeutics for JCAR017 (under review by the US Food and Drug Administration). This study shows that acquisition agreements led by the major pharmaceutical companies with amounts ranging from $0.1 billion to $74 billion quickly increased over the years but reached a peak in 2019 with more than $153 billion in total value ($15.3 billion in average value). This was heavily weighted by the two mega-agreements in 2019: Bristol-Myers Squibb acquired Celgene for $74 billion, and Takeda acquired Shire for $62 billion.

Before agreeing to a buyout, it is not uncommon for a major pharmaceutical company to first establish a partnership with a biotechnology company to validate its technology and eliminate risk to the assets of the pharmaceutical company (208). For example, Astellas Pharma forged a licensing agreement with Universal Cells in October 2017 to utilize its proprietary Universal Donor Cell technology, and subsequently announced the acquisition of Universal Cells in February 2018. Such a partnership-before-acquisition model was also seen when Gilead Sciences (and its subsidiary Kite Pharma) acquired Cell Design Labs and Takeda acquired TiGenix. Another model is partnership with option to acquire, where the major pharmaceutical company is granted the right to acquire the biotechnology

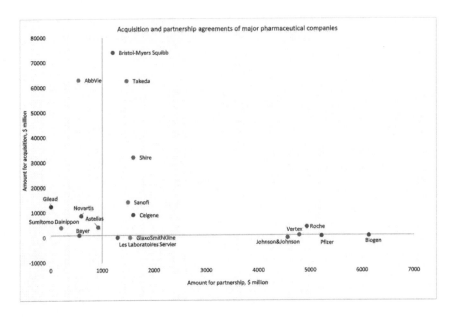

FIGURE 5.15 Preference towards partnerships and acquisitions of big pharma.

company after certain milestones have been reached. For example, Pfizer retained the right to acquire Vivet Therapeutics if the phase 1/2 trial of VTX-801 confirms its clinical promise for the treatment of Wilson disease.

Some of the major pharmaceutical companies show clear preferences towards acquisition or partnership (Figure 5.15). For example, Gilead Sciences was rarely engaged in the partnership agreements of RMs but immediately gained a range of RM assets upon acquisition of Kite Pharma in August 2017, whereas Pfizer could be seen as a company that made more investments in partnership agreements than in outright acquisitions. Merck could be a slightly different case: as a member of only a few manufacturers that have an industrialized process to make viral vectors (209), Merck is engaged not only as a licensee but also as a provider of manufacturing services. For example, Merck has contracted with Bluebird Bio to manufacture the lentiviral vectors.

5.3.3 FUTURE TRENDS AND STRATEGIES

- **Internal and External Collaboration**

Undoubtedly, the partnerships for RMs will continue to be multidimensional collaborations to overcome all relevant challenges faced in the market access of RMs. Internally, traditional linear commercialization processes involving a series of handoffs, starting with early research and ending with commercial launch, may not suffice for RMs (210). The functional boundaries between separate company departments may become less obvious, considering that all key processes require simultaneous input from the quality assurance and quality control sectors. Cooperation between the manufacturing and the supply chain teams is important, as is communication so that these teams can be integrated with the commercial teams. This level of cooperation is fairly unique for biologicals, and in particular, for RMs (210).

External, the materialization of the commercial potential of RMs will rely on an "innovative ecosystem," which was proposed by Adner (211) as a collaborative arrangement through which each individual company combines its individual experiences into a coherent solution. The core element of an innovative ecosystem is that it leverages synergies across multiple companies/organizations to bring value to customers that no company/organization alone could deliver (212). Academia, not-for-profit organizations (e.g., patient group or charities), biotechnology companies, and service companies will continue to intricately collaborate in both upstream and downstream activities, spanning from early research to achieving developmental milestones, which will promote the value of assets and attract well-financed partners.

- **Emerging Field for Optimization and Innovation**

When examining agreements with values greater than $100 million, CAR-T and T-cell receptor therapies remained the main assets that triggered most investments. One promising business for next-generation CAR-T is using gene-editing technology to develop allogeneic CAR-T cell products that can be used on an "off-the-shelf" basis. The advantages of allogeneic CAR-T cell products include the elimination of complex manufacturing/logistics processes required with autologous products, increased scalability, greater product consistency, and increased access for patients who are ineligible for autologous products due to T-cell harvesting failures (213) (Table 5.2). Notably, for the development of allogeneic CAR-T cell therapy, Sangamo Therapeutics has partnered with Kite Pharma

TABLE 5.2
Comparison Between Allogeneic Products and Autologous Products

Allogeneic Approach	Autologous Approach
Immunological Issues	
• High risk of immune rejection	• Low risk of immune rejection
• Cells more likely to survive only transiently, so stimulation of the production of host paracrine factors may play a bigger role	• May be the preferable option when more permanent cell integration/differentiation is required
Patient-centric Factors	
• Stem cells characteristically from young, healthy adult donors, so high in quality and number; donors chosen specifically for their potential ability to yield a treatment	• Stem cells from older patients both less likely to yield therapy easily (due to lower proliferative capacity) and for that therapy to have a very large therapeutic benefit
Commercial-Scale Manufacturing	
• Therapy is derived from non-self donor cells via the creation of a cell bank system	• Therapy is derived from a patient's own cells
• Each batch consists of up to 2,000 doses	• Each dose is typically a batch in itself
• Lower cost of goods (compared to autologous)	• High cost of goods
• Mass manufacturing fits in well with present manufacturing practices at biopharma companies	• Personalized manufacturing does not fit in well with present manufacturing practices at biopharma companies
• Automated, large-scale manufacturing easier to achieve	• Automated, large-scale manufacturing challenging due to the inconsistency in patients' cells in terms of culturing properties and due to the risk of cross contamination
• Easier to scale up manufacturing due to product uniformity (because it is produced from a cell bank)	• Challenging to scale up manufacturing due to the variability in different patients' cells
• If product is damaged/lost, patient can receive another identical dose quickly	• If product is damaged/lost, patient must endure significant delay while another dose is produced, so a high level of product security/quality control is needed

TABLE 5.2 *(Continued)*
Comparison Between Allogeneic Products and Autologous Products

Allogeneic Approach	Autologous Approach
Business Models	
• Functions on an "off-the-shelf" business model	• Functions on a "service-based" business model
• Fits in well with current biopharma model	• Does not fit in with current biopharma model
• More commercially attractive, as it can be used in both acute and chronic disease settings (as it is available for use relatively quickly)	• Less commercially attractive, as it can only be used in the nonacute disease setting (due to time lag for therapy to be manufactured)
• Well suited to high-incidence/prevalence disease, as a large number of doses can be produced much more easily and cheaply due to high scalability; also, can enter niche areas	• Well suited to orphan/niche indications; in such indications, autologous cell therapy can first achieve a high reimbursement price to recuperate its high cost of goods and so enjoy an acceptable gross profit margin, and second it can compete with allogeneic products, as economies of scale achieved with the allogeneic approach are much less relevant in small-volume therapeutic areas
Reimbursement Price Potential	
• Expected to secure high reimbursement price, especially for orphan/niche/life-threatening indications; likely to be commercially pursued for both rare and more common diseases	• Expected to secure high reimbursement price, especially for orphan/niche/life-threatening indications; likely to gravitate toward orphan/niche diseases in the long term

Source: Malik Nafees, et al. Cell Therapy Landscape: Autologous and Allogeneic Approaches. Book: Translational Regenerative Medicine (pp. 87–106). 2015. DOI: 10.1016/B978–0-12–410396–2.00007–4

(total amount of $3.16 billion) utilizing its ZFN platform, Cellectis has partnered with Servier (total amount of $930 million) utilizing its transcription activator-like effector nuclease (TALEN) gene-editing technology. Next-generation CAR-T cell therapies also centered on the improvement of the safety profile by incorporating "on/off switch" small molecules to control the timing of T-cell activation or to modulate cytolytic activity (214). For example, Throttle is an "on switch" technology platform from Cell Design Labs that has been an important contributor to the acquisition of that company by Kite Pharma in 2017.

Apart from CAR-T technology, gene-editing technology is also being leveraged in other therapeutic areas (Figure 5.16). For example, CRISPR Therapeutics is eligible to receive a total amount of $2.63 billion from Vertex for the use of its gene-editing technology (CRISPR-Cas9) for cystic fibrosis and sickle cell disease. Another innovative area is focused on the novel vector to improve potency, evade neutralizing antibodies to vectors, and improve tropism to certain

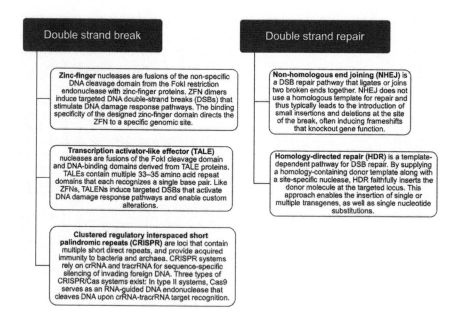

FIGURE 5.16   Gene editing technology: mechanism and advances.

Source: Li HY, et al. Applications of genome editing technology in the targeted therapy of human diseases: mechanisms, advances and prospects. Nature. 2020. Open access paper, no need to license for reuse

tissues that include tissues in the CNS. For example, Voyager Therapeutics has collaborated with AbbVie to develop vectorized antibodies for indications that include Parkinson disease, Alzheimer disease, and other neurodegenerative diseases. Additionally, the field of R&D has noticed the potential of using artificial intelligence (AI) to develop novel AAV capsids with optimized delivery. Dyno Therapeutics, founded in 2018 as a spin-off of Harvard University, has an AI-based platform called CapsidMap. In October 2018, the company signed its third collaboration agreement with Roche worth up to $1.8 billion to develop AAV used for CNS disease, after its agreements signed with Novartis and Sarepta Therapeutics in May 2020. The implication is that much room still exists for R&D for RMs in terms of product optimization and innovation.

- **The Future for Academia and Biopharmaceutical Companies**

Apart from partnership and acquisition, some major pharmaceutical companies are shifting their focus to enhance internal R&D and manufacturing capacities. The capacity for viral vector and cell manufacturing has been constrained, and developing in-house manufacturing capabilities can be viewed as a competitive advantage (138). From our observations, Novartis, Pfizer, Sanofi, GlaxoSmithKline, Gilead Sciences, Vertex, and Astellas Pharma have announced the expansion of the manufacturing facilitates for RMs in the past 2 years for their already approved products or existing pipeline products. However, it may become more typical to see collaborations between the major pharmaceutical companies because first, some of the pioneering biotechnology companies (e.g., Kite Pharma and Spark Therapeutics) now have been acquired by the major pharmaceutical companies and, second, biotechnology companies have begun to explore the possibility of combining of CAR-T cell therapies with immune checkpoint inhibitors (ICIs) to combat solid tumors (215), while all approved ICIs are dominated by the major pharmaceutical companies (216). If the trend for the acquisition of biotechnology companies continues, ultimately all of the key RM assets will be held by the major pharmaceutical companies. Thus, more collaborations between large asset owners to jointly investigate the potential of other combination regimens consisting of RMs will occur.

Biotechnology companies have had access to unprecedented levels of capital via VC, IPO routes, and equity investment. This has been true, even during the period of the COVID-19 pandemic (169). In the future, biotechnology companies may take an increasing share of the market from the major pharmaceutical companies by developing and commercializing products independently (217). For example, uniQure, which was perceived as a potentially attractive target for a buyout by the major pharmaceutical companies, has sold its lead hemophilia B candidate to CSL Behring, making it less likely to be acquired. This move was made to reprioritize the received investment capital to other programs (e.g., AMT-130 for Huntington disease). However, uniQure claims that it still aspires to be a fully vertically integrated commercial entity and plans to bring other gene therapies to the market on its own (218). Likewise, academia may gain more autonomy in the future, considering that it may use capital raised by an IPO or reinvest royalties and milestone payments to further its own prioritized assets. The development of academic discovery through spin-offs may be significantly more profitable than stopping the development of a very early asset. For example, the University of Texas spun off Taysha Gene Therapeutics and has expressed a willingness to build a sustainable company, changing the status quo where academia spin-off biotechnology companies (e.g., Spark Therapeutics and Nightstar Therapeutics) primarily were purchased by the major pharmaceutical companies.

In the long run, it may be more difficult for the major pharmaceutical companies to buy innovation with more determined benefits than in the past. They will have to share more risk and engage earlier to ensure their future ownership of the discoveries. Otherwise, the major pharmaceutical companies need to continue to increase the price they pay for these assets but still face an uncertain future regarding the return on investment. Additionally, highly specialized services for RMs currently can be provided only by scattered, small-to-medium-sized contract organizations. This fact will likely whet the appetite of large contract manufacturing organizations (CMOs) for the acquisition of these companies to expand the presence of CMOs in the RMs field, such as the acquisition of Brammer Bio by Thermo Fisher in March 2019. The result may be less opportunity for expertise sharing, which may make it difficult for biotechnology companies to partner with small service companies due to the monetization of the service offering.

5.4   CASE ANALYSES: PARTNERSHIPS FOR APPROVED RMS

5.4.1   RMs Commercialized on the Market

5.4.1.1   Alofisel

- Discovery and pre-clinical stage
  - *Spin-off from academia*: TiGenix was founded in February 2000 by Dr. Frank P. Luyten and Gil Beyen as a spin-off from the Katholieke Universiteit Leuven and the Universiteit Gent.
- Clinical trials
  - *Manufacturing contract*: Lonza and TiGenix announced an agreement for the supply of TiGenix's eASC product, Cx601, in February 2015. Lonza will manufacture material for the phase 3 trial of Cx601 in the United States at Lonza's cell therapy production facility in Walkersville, Maryland.
- Prior to market authorization
  - *Exclusive licensing and development*: Takeda Pharmaceutical and TiGenix NV announced an exclusive licensing and development deal for Cx601 to treat complex perianal fistulas in patients with Crohn disease in December 2016, under which Takeda acquired exclusive rights to develop and commercialize Alofisel for complex perianal fistulas in Crohn patients outside of the United States. TiGenix will maintain the rights to develop the compound for new indications, and will hold onto US rights. TiGenix has submitted a marketing authorization application to the European Medicines Agency (EMA) for the compound recently based on 24-week data from the ADMIRE-CD phase 3 clinical trial, which lasted 24 weeks.
  - *Exclusive licensing and commercialization*: TiGenix has received a confirmation notification from Takeda Pharmaceuticals stating that it has decided to exercise its option to develop and commercialize Cx601 in both Japan and Canada in December 2016.
  - *Exclusive licensing:* Mesoblast has granted TiGenix exclusive access to certain of its patents to support global commercialization of the adipose-derived MSC product Cx601 for the local treatment of fistulae in December 2017. The agreement includes the right for TiGenix to grant sublicenses to affiliates and third parties, including TiGenix's current development and commercialization partner that is outside of United States.
- Market authorization
  - *Favorable decisions for market authorization*: The Committee for Medicinal Products for Human Use (CHMP) of the EMA adopted a positive opinion recommending marketing authorization for Cx601 for the treatment of complex perianal fistulas in Crohn disease, one of the most disabling manifestations of the disease, in December 2017.

- *Market authorization:* Cx601 received market authorization for the treatment of complex perianal fistulas in adult patients with non-active/mildly active luminal Crohn disease from EMA in on 23 March 2018.
- Acquisition
  - *Intention to acquire*: Takeda announced their intention to acquire TiGenix in January 2018, after the positive opinion was delivered by CHMP.
  - *Acquisition completed:* Takeda completed its acquisition of TiGenix following expiration of the squeeze-out period in July 2018, TiGenix became a wholly owned subsidiary of Takeda (Figure 5.17).

5.4.1.2 Imlygic

- Discovery or pre-clinical
  - *Spin-off from academia*: BioVex was launched in 1999 as a spin-off company of University College London to exploit research undertaken by David Latchman at the UCL Medical Molecular Biology Unit, Department of Biochemistry.
- Early stage of clinical trial
  - *Phase 1/2 clinical trial*: In 2005, the phase 1/2 clinical trial of OncoVEXGM-CSF showed that it has been well tolerated and

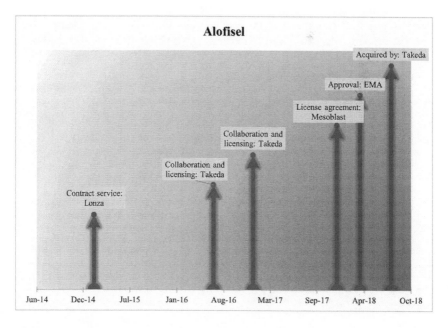

FIGURE 5.17 Partnership agreements for Alofisel.

evidence of biological activity observed, including virus replication, tumor necrosis, and granulocyte macrophage colony-stimulating factor (GM-CSF) expression.

- *Phase 2 clinical trial*: In 2009, the completed phase 2 clinical trial reported a high objective durable response rate and a high complete response rate, combined with a relatively benign side effect profile of OncoVEX (GM-CSF) for the treatment of advanced melanoma.

- Acquisition
  - *Acquisition*: Amgen announced that the companies have entered into a definitive acquisition agreement under which Amgen has agreed to acquire BioVex Group. BioVex is developing OncoVEX (GM-CSF), a novel oncolytic vaccine in phase 3 clinical development, that may represent a new approach to treating melanoma and head and neck cancer.

- Clinical trials extension
  - *Combination with anti–PD-1 immunotherapy*: Amgen and Merck announced in February 2014 that they had entered into an agreement to evaluate the safety and efficacy of talimogene laherparepvec, an investigational oncolytic immunotherapy, combined with MK-3475, an investigational anti–PD-1 immunotherapy, in a phase 1b/2 study of patients with mid- to late-stage melanoma.
  - *Combination with anti–PD-1 immunotherapy*: Amgen and Merck announced an expanded collaboration to evaluate the efficacy and safety of talimogene laherparepvec, Amgen's investigational oncolytic immunotherapy, in combination with Keytruda (pembrolizumab), Merck's anti–PD-1 therapy, in a phase 1, open-label trial of patients with recurrent or metastatic squamous cell carcinoma of the head and neck (SCCHN).
  - *Combination with anti–PD-1 immunotherapy*: Amgen announced a collaboration with Roche on a phase 1b study to evaluate the safety and efficacy of Imlygic in combination with Roche's investigational anti–PD-1 therapy, atezolizumab (also known as MPDL3280A), in patients with triple-negative breast cancer and colorectal cancer with liver metastases.

- Market authorization
  - *Market authorization in the United States*: Imlygic received market authorization from the FDA in October 2015 for the treatment of unresectable cutaneous, subcutaneous, and nodal lesions in patients with melanoma recurrent after initial surgery.
  - *Market authorization in the EU*: Imlygic received market authorization from the EMA in December 2015 for the same indication as the FDA has granted (Figure 5.18).

5.4.1.3 Kymriah

- Discovery and pre-clinical stage
  - *Exclusive licensing:* Exclusive licensing agreements were reached between Novartis and the University of Pennsylvania in August 2012.

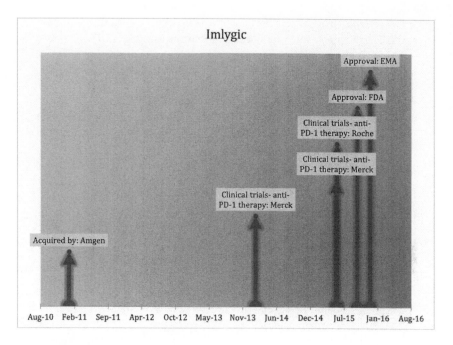

FIGURE 5.18   Partnership agreements for Imlygic.

Under the terms of the agreement, Penn grants Novartis an exclusive worldwide license to the technologies used in an ongoing trial of patients with chronic lymphocytic leukemia and future CAR-based therapies developed through the collaboration. Novartis will invest in the establishment of the Center for Advanced Cellular Therapies (CACT) and future research of the technology. Additional milestone and royalty payments to Penn are also part of the agreement.

- Clinical trials
  - *Manufacturing contract:* Oxford BioMedica announced in May 2013 that it has signed an agreement with Novartis to manufacture clinical-grade material utilizing Oxford BioMedica's LentiVector gene delivery technology. Under the terms of the agreement, Oxford BioMedica will be responsible for manufacturing several batches of a lentiviral vector encoding CTL019 technology. This manufacturing contract was extended in October 2014 and July 2017, respectively.
  - *Manufacturing contract:* Fraunhofer Institute for Cell Therapy and Immunology (IZI) in Germany collaborated with Novartis on the technology transfer of an investigational chimeric antigen receptor therapy (CTL019) in January 2015. It aimed to obtain the specific manufacturing authorization in accordance with Section 13 of the

German Drug Act. Subsequently the Fraunhofer Institute will assist in clinical programs for patients in Europe using CTL019.

- Market authorization
  - *Market authorization by the FDA:* Kymriah received market authorization from the FDA for the treatment of B-cell precursor acute lymphoblastic leukemia (ALL) in August 2017.
  - *Extension of indication by the FDA:* Kymriah was approved for the extension of indication for diffuse large B-cell lymphoma (DLBCL) in May 2018.
  - *Market authorization by the EMA:* Kymriah received market authorization from the EMA for the treatment of ALL and DLBCL in August 2018.
- Post-market authorization
  - *Manufacturing contract:* The IZI and Novartis announced in August 2018 that a further agreement has been concluded between them for the manufacture of the CAR-T cell therapy for patients in Europe over the next few years.
  - *Manufacturing contract:* Cellular Biomedicine (CBMG) has entered into a strategic licensing and collaboration agreement with Novartis to manufacture and supply the CAR-T cell therapy Kymriah (tisagenlecleucel) in China in September 2018. Novartis will be the exclusive holder of the marketing license. CBMG will take the lead in the manufacturing process, and Novartis will lead distribution, regulatory, and commercialization efforts in China.

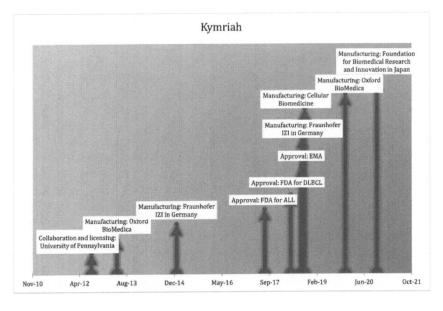

FIGURE 5.19   Partnership agreements for Kymriah.

- *Manufacturing contract:* Oxford BioMedica announced in December 2019 that it has extended its commercial supply agreement with Novartis for the manufacture of lentiviral vectors for the Novartis CAR-T portfolio, including five lentiviral vectors for CAR-T products. This builds on the existing 3-year commercial supply agreement signed by the parties in July 2017.
- *Manufacturing contract:* Novartis announced the receipt of marketing authorization from Japan's Ministry of Health, Labor, and Welfare (MHLW) for the Foundation for Biomedical Research and Innovation at Kobe (FBRI) to manufacture and supply commercial Kymriah (tisagenlecleucel) for patients in Japan in October 2020. This approval makes FBRI the first and only approved commercial manufacturing site for CAR-T cell therapy in Asia (Figure 5.19).

5.4.1.4   Luxturna

- Clinical trials
  - *Before foundation*: Voretigene neparvovec received orphan drug designation for Leber congenital amaurosis (LCA) from the EMA in April 2012. Phase 3 clinical trials for voretigene neparvovec were initiated in November 2012. The intervention was given either at the Children's Hospital of Philadelphia (CHOP) or University of Iowa.
  - *Spin-off from academia:* Spark was founded in 2013 as a spin-off company of CHOP. In October 2013, Spark and CHOP entered into a technology and license agreement for certain commercialization licenses to be provided to Spark so that it could execute on a plan to develop services, methods, and marketable products for commercialization. The license agreement required that Spark reimburse CHOP for the patent costs related to underlying licensed rights incurred post-deal closure.
  - *Exclusive licensing agreements*: Spark entered into a licensing agreement with the University of Pennsylvania (Penn) in December 2014, under which Penn granted Spark an exclusive, worldwide license, with the right to sublicense, to certain patent rights owned by Penn related to certain pro-viral plasmids that are useful in the manufacture of certain gene therapy products for the treatment of choroideremia. Under the terms of the license agreement, we are obligated to use commercially reasonable efforts to develop and commercialize licensed products.
  - *Option licensing agreement*: Spark Therapeutics and Clearside Biomedical entered into an option agreement in April 2015, under which Spark acquired exclusive rights to license Clearside's microinjector technology to deliver gene therapies to the back of the eye. The companies will explore the feasibility of using Clearside's microinjector technology to deliver viral vectors to the choroid and the retina through the suprachoroidal space (SCS). In February 2016, the initial study was completed, and Spark elected not to extend the arrangement or license the technology, which terminated the Spark agreement in accordance with its terms.

- Market authorization
  - *Market authorization in the United States*: Luxturna was approved by the FDA for the treatment of patients with confirmed biallelic RPE65 mutation-associated retinal dystrophy in December 2017.
  - *Market authorization in the EU*: Luxturna was approved by the EMA for the same indication in July 2018.
- Post-market authorization
  - *Distribution and commercialization*: Novartis announced a licensing agreement with Spark Therapeutics covering development, registration, and commercialization rights to voretigene neparvovec in markets outside the United States in January 2018. Spark Therapeutics retains exclusive rights for Luxturna in the United States and will retain responsibility for obtaining EMA approval. Spark Therapeutics will be responsible for the supply of voretigene neparvovec worldwide under a separate manufacturing and supply agreement with Novartis.
- Acquisition
  - Roche and Spark Therapeutics announced the completion of the acquisition following the receipt of regulatory approval from all government authorities required by the merger agreement in December 2019. Spark will continue its operations in Philadelphia as an independent company within the Roche Group (Figure 5.20).

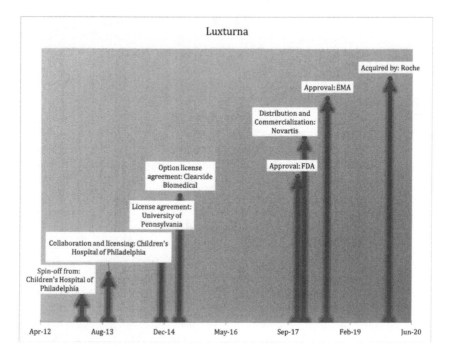

FIGURE 5.20 Partnership agreements for Luxturna.

5.4.1.5   Strimvelis

- Clinical trials
  - *Exclusive licensing agreements*: GlaxoSmithKline (GSK), Fondazione Telethon, and Fondazione San Raffaele announced a new strategic alliance to research and develop novel treatments to address rare genetic disorders using gene therapy carried out on stem cells taken from the patient's bone marrow (ex vivo) in October 2010. Under the terms of the agreement, GSK will gain an exclusive license to develop and commercialize an investigational gene therapy for adenosine deaminase deficiency (ADA)–related severe combined immune deficiency (ADA-SCID). Phase 1/2 studies have demonstrated the potential of this treatment option to restore long-term immune function and protect against severe infections in children with ADA.
  - *Manufacturing contract*: GSK signed an agreement with MolMed in August 2011. As part of the agreement, MolMed will provide its expertise in process development and its manufacturing competencies and capacity for the production of viral vectors and cell transduction for gene therapy for ADA-SCID. This agreement was extended in 2013 and 2015, respectively.
- Market authorization
  - *Market authorization in the EU*: Strimvelis was approved by the EMA for the treatment of ADA-SCID in May 2016.
- Post-market authorization
  - *Asset purchase*: GSK and Orchard Therapeutics announced a strategic agreement under which GSK will transfer its portfolio of Strimvelis to Orchard, securing the continued development of the program and access for patients. GSK will continue to invest in the development of its platform capabilities in cell and gene therapies, with a focus on oncology. GSK will become an investor in Orchard Therapeutics, receiving a 19.9% equity stake along with a seat on the company's board. GSK and Orchard will exchange manufacturing, technical, and commercial insights and learnings on the development of gene therapy medicines to ensure the success of the assets (Figure 5.21).

5.4.1.6   Yescarta

- Discovery and pre-clinical stage
  - *Exclusive licensing*: Axicabtagene was initially developed at National Cancer Institutes (NCI) by Steven Rosenberg, of the Surgery Branch in NCI's Center for Cancer Research (CCR), and his colleagues. It was later licensed to Kite Pharma in April 2013 for further development and commercialization.
- Clinical trials
  - *Combination with PD-L1 inhibitor*: Kite Pharma announced in March 2016 that it has entered into a clinical trial collaboration with Genentech, a member of the Roche Group, to evaluate the safety and

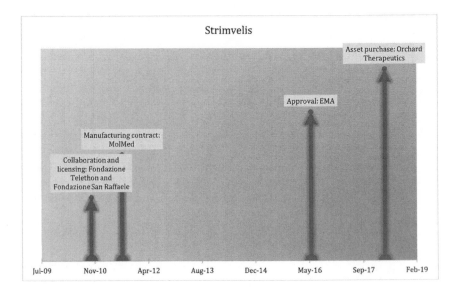

FIGURE 5.21   Partnership agreements for Strimvelis.

efficacy of KTE-C19, in combination with atezolizumab (also known as MPDL3280A), in patients with refractory, aggressive non-Hodgkin lymphoma (NHL).

- *Logistic chain service*: Kite Pharma and Vitruvian Networks announced a strategic partnership to create a software solution to support commercial availability of T-cell therapies in December 2016. Together, the parties will design and develop a platform for patients, physicians, and treatment centers that enables commercial-scale ordering, logistics, monitoring, and delivery of autologous cell therapies if they are FDA-approved, including axicabtagene ciloleucel (formerly known as KTE-C19), Kite's lead investigational engineered T-cell therapy for aggressive NHL.

- *BLA submission*: In December 2016, Kite Pharma announced that it has submitted to the US Food and Drug Administration (FDA) a Biologics License Application (BLA) for KTE-C19 as a treatment for patients with refractory aggressive B-cell NHL.

- *Distribution and commercialization*: Kite Pharma announced in January 2017 that it has signed a strategic partnership deal with Japan's Daiichi Sankyo Co. for axicabtagene ciloleucel in Japan. Under the terms of the deal, Daiichi Sankyo will take over development and commercialization of the compound in Japan. Kite will provide specific technical transfer services for Daiichi Sankyo.

- *Joint venture agreement*: Shanghai Fosun Pharmaceutical announced the establishment of a Sino-foreign cooperative enterprise in China through its wholly owned subsidiary to introduce KTE-C19, CAR-T products of Kite Pharma, to benefit lymphoma patients with world-leading therapy in January 2017. The cooperative enterprise will be registered in Shanghai and owned equally between Fosun Pharma and Kite Pharma. Under the terms of the agreement, Fosun Pharma intends to contribute US$20 million, and Kite Pharma contributes the right to exclusive use of its products and proprietary technology at the value of US$20 million. Each of them owns 50% equity interests in the cooperative enterprise.

- Acquisition
  - Gilead Sciences announced the acquisition of Kite Pharma for $11.9 billion in August 2017. This immediately positions Gilead as a leader in cell therapy. Kite's lead CAR-T therapy candidate, axicabtagene ciloleucel, was under priority review in the United States and expedited review in the EU.
- Market authorization
  - *Market authorization by the FDA:* Yescarta received market authorization from the FDA for the treatment of DLBCL and primary mediastinal large B-cell lymphoma (PMBCL) in October 2017.
  - *Market authorization by the EMA:* Yescarta received market authorization from the EMA for the treatment of the same indication as the FDA allowed.
- Post-market authorization
  - *Combination with a monoclonal antibody for improving efficacy*: Kite has entered into a clinical trial collaboration with Pfizer to evaluate the safety and efficacy of the investigational combination of Yescarta and Pfizer's utomilumab, a fully humanized 4–1BB agonist monoclonal antibody, in patients with refractory large B-cell lymphoma in January 2018.
  - *Combination with a monoclonal antibody for improving safety*: Kite and Humanigen announced the formation of a clinical collaboration to conduct a phase 1/2 study of lenzilumab, an investigational anti–GM-CSF monoclonal antibody, with Yescarta in patients with relapsed or refractory DLBCL in May 2019. The objective of this study is to determine the effect of lenzilumab on the safety of Yescarta.
  - *Combination with a monoclonal antibody for improving safety*: Kite and Kiniksa Pharmaceuticals announced that the companies have entered into a clinical collaboration to conduct a phase 2, multicenter study of mavrilimumab, an investigational anti–GM-CSF monoclonal antibody, in combination with Yescarta in patients with relapsed or refractory DLBCL in December 2019. The objective of the study is to determine the effect of mavrilimumab on the safety of Yescarta (Figure 5.22).

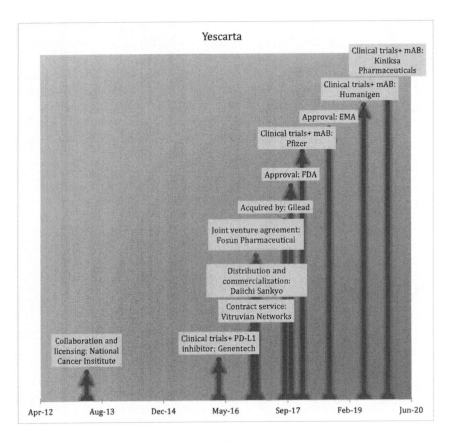

FIGURE 5.22   Partnership agreements for Yescarta.

5.4.1.7 Zolgensma

- Discovery and pre-clinical stage
  - *Exclusive licensing*: In February 2009, REGENXBIO entered into a license agreement with the University of Pennsylvania (Penn) for exclusive, worldwide rights to certain patents owned by Penn underlying the company's NAV Technology Platform. Under the terms of the agreement, in consideration for the license, the company issued to Penn 24.5% of the then outstanding membership interest in the company on a fully diluted basis after issuance. REGENXBIO was obligated to pay Penn royalties on net sales and sublicense fees, if any.
  - *Exclusive licensing*: In October 2013, AveXis entered into an exclusive license agreement with Nationwide Children's Hospital (NCH). Under the terms of the agreement, NCH granted AveXis an exclusive, non-transferable, worldwide license to certain patents held by NCH for the therapy and treatment of spinal muscular atrophy (SMA).

- *Exclusive licensing*: In March 2014, REGENX Biosciences and AveXis have entered into an exclusive agreement for the development and commercialization of products to treat SMA using NAV rAAV9 vectors. REGENX granted AveXis an exclusive, worldwide license, with rights to sublicense, to REGENX's NAV rAAV9 vector for the treatment of SMA disease in humans.
- Clinical trial
  - *Exclusive licensing*: Asklepios BioPharmaceutical announced a license agreement in June 2015, which granted AveXis rights to AskBio's proprietary self-complementary (SC) technology, also known as duplex vectors. This technology was utilized in AveXis's phase 1 AAV gene transfer clinical trial in SMA1 patients.
  - *Exclusive licensing*: AveXis and Genethon entered into an exclusive, worldwide license agreement for in vivo gene therapy delivery of the AAV9 vector into the CNS for the treatment of SMA. Genethon granted AveXis a license to patents in the United States, Europe, and Japan, for the AAV9 SMN product and in vivo gene therapy delivery of the AAV9 vector into the CNS using intrathecal or intravenous routes of administration for the treatment of SMA.
- Acquisition
  - Novartis announced in April 2018 that it had acquired AveXis for $218 per share or a total of $8.7 billion in cash. If approved, AVXS-101 would be a first-in-class one-time therapy that addresses the root genetic cause of SMA by effectively replacing the defective *SMN1* gene. In a clinical study, AVXS-101 showed lifesaving efficacy, with all 15 infants treated event-free at 20 months compared with an event-free survival rate of 8% in an historical cohort. Novartis announced in December 2018 that the FDA had accepted the company's BLA for AVXS-101.
- Prior to market authorization
  - The Novartis subsidiary AveXis expanded its gene therapy manufacturing capacity in April 2019 by agreeing to purchase AstraZeneca's campus in Longmont, Colorado, which has nearly 700,000 square feet of space for biologic drug manufacturing as well as offices, laboratories, warehousing, and utilities.
- Market authorization
  - *Market authorization by the FDA:* Zolgensma received market authorization from the FDA for the treatment of pediatric patients less than 2 years old with biallelic mutations in the *SMN1* gene in May 2019.
  - *Market authorization by the EMA:* Kymriah received market authorization from the EMA in May 2020 for the same indication as the FDA.
- Post-market authorization
  - *Manufacturing contract*: BIA Separations announced in July 2019 that it was collaborating with AveXis to enhance the commercial

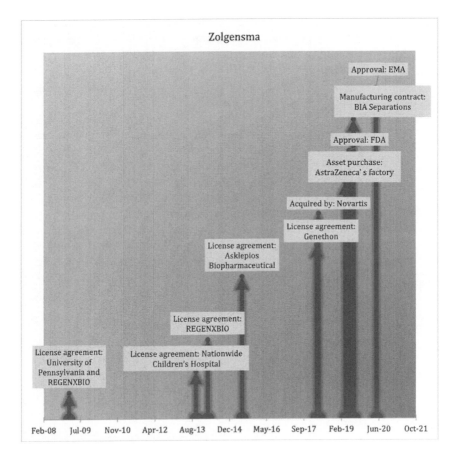

FIGURE 5.23 Partnership agreements for Zolgensma.

purification process for its gene therapy pipeline, beginning with Zolgensma (onasemnogene abeparvovec-xioi). BIA Separations will provide expertise in next-generation biomolecule purification processes and long-term supply of CIMmultus monolithic columns. BIA was selected as a result of its expertise in viral and DNA purification, particularly AAV, and chromatographic technology for the purification and analysis of biomolecules (Figure 5.23).

5.4.1.8 Zynteglo

- Discovery, pre-clinical, and early clinical stage
  - The development of betibeglogene autotemcel was led by Marina Cavazzana (director of the Biotherapy Department at Necker-Enfants Malades Hospital) and Philippe Leboulch (head of the Institute of Emerging Diseases and Innovative Therapies of the CEA and INSERM; professor of medicine, University of Paris; and visiting

professor, Harvard Medical School). Professor Leboulch and Irving London founded Genetix Pharmaceuticals in April 1992, and the company was renamed Bluebird Bio in September 2010.

- In September 2010, one publication in the journal *Nature* highlighted the positive results of LentiGlobin gene therapy treatment in a young adult with severe β-thalassemia. The patient, who had been transfusion dependent since early childhood, has become transfusion independent for the past 21 months—more than 2 years after treatment with the LentiGlobin vector.
- Late clinical stage
  - *Manufacturing contract*: Lonza and Bluebird Bio entered into a strategic manufacturing agreement providing for the future commercial production of Bluebird Bio's Lenti-D and LentiGlobin drug products in June 2016. This agreement follows a successful multiyear clinical manufacturing relationship and provides Bluebird Bio with a path to commercial supply, including dedicated production suites within Lonza's state-of-the-art facility. This facility is currently under construction for the clinical and commercial supply of viral vectors and virally modified cell therapy products. Under this multiyear agreement, Lonza will complete the suite design, construction, and validation along with process validation, prior to anticipated commercial launch.
  - *Manufacturing contract*: Bluebird Bio and Apceth Biopharma entered into a strategic manufacturing agreement in December 2016 providing for the future European commercial production of Bluebird Bio's LentiGlobin product candidate for transfusion-dependent β-thalassemia. Apceth Biopharma will perform clinical manufacturing, process validation activities, and commercial manufacturing for LentiGlobin and Lenti-D drug products to support the treatment of European patients with transfusion-dependent β-thalassemia and cerebral adrenoleukodystrophy, respectively.
  - *Value-based payment*: Bluebird Bio participated in a consortium led by Duke University's Robert J. Margolis, MD, Center for Health Policy, to develop a broadly supported path for value-based payment reform models for gene therapies and other innovative treatments. The consortium that is composed of patient advocates, payers, manufacturers, and providers, as well as experts in regulatory science, law, and policy, will collaborate to outline a path forward for payment reform involving innovative therapies, including genomic medicines for rare diseases.
  - *Manufacturing contract*: Merck announced in December 2017 that it signed a commercial supply agreement to manufacture viral vectors for Bluebird Bio for its use in potentially transformative gene therapies. Under the multiyear agreement, the Life Science business sector of Merck will manufacture lentiviral vectors for Bluebird Bio's drug products developed to treat a variety of rare genetic diseases. Bluebird Bio is a clinical-stage company that develops potentially

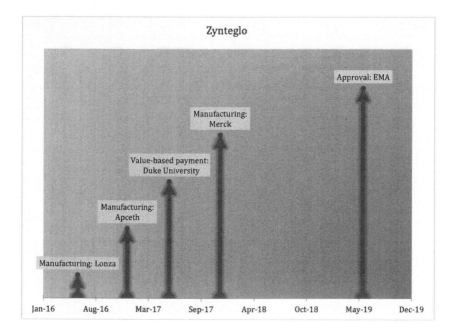

FIGURE 5.24   Partnership agreements for Zynteglo.

transformative gene and cell therapies for severe genetic diseases and T-cell–based immunotherapies for cancer.

* Market authorization
  * *Market authorization by the EMA*: Zynteglo received market authorization from the EMA in May 2019 for the treatment of β-thalassemia in patients 12 years and older who require regular blood transfusions (Figure 5.24).

5.4.2   RMs Withdrawn from the Market

5.4.2.1   Glybera

* Discovery and pre-clinical stage
  * *Exclusive licensing*: Together with collaborators from the University of British Columbia (UBC), Xenon Pharmaceuticals demonstrated that the LPLS447X variant resulted in increased lipoprotein lipase (LPL) enzyme activity leading to reduced triglyceride levels in humans. Effective August 2000, Xenon Pharmaceuticals entered into a sublicense and research agreement with uniQure (formerly Amsterdam Molecular Therapeutics), pursuant to which uniQure was granted an exclusive, worldwide sublicense under certain intellectual property controlled by Xenon Pharmaceuticals to develop and commercialize technology and compounds related to LPLS447X.

- Market authorization
  - *Market authorization*: Glybera received market authorization from the EMA as an indication for adult patients diagnosed with familial lipoprotein lipase deficiency (LPLD) and suffering from severe or multiple pancreatitis attacks despite dietary fat restrictions in October 2012.
- Post-market authorization
  - *Distribution and commercialization*: uniQure gave Chiesi exclusive rights to commercialize Glybera, the first gene therapy product approved in the EU for the treatment of the orphan disease LPLD, for which there is currently no treatment, as well as for uniQure's pipeline product for hemophilia B, in Europe and selected other countries (Brazil, Mexico, Pakistan, Turkey, Russia, and the Commonwealth of Independent States [CIS] countries, plus China for Glybera only) in July 2013.
  - *Distribution and commercialization*: uniQure announced an exclusive distribution agreement in May 2014, under which Medison will market Glybera, uniQure's gene therapy product for the treatment of LPLD, in Israel and the Palestinian Authority. Medison will also be responsible for obtaining regulatory approval for Glybera in both territories.
- Market withdrawal
  - *Not renewal of the market authorization*: uniQure announced in April 2017 that it would not pursue renewal of the Glybera (alipogene tiparvovec) marketing authorization in Europe when it was scheduled to expire on 25 October 2017. The decision to not pursue marketing authorization renewal of Glybera in Europe involved a thoughtful and careful evaluation of patient needs and the clinical use of the therapy and is not related to any risk-benefit concern.
  - *Agreement termination*: uniQure announced in July 2017 that it entered into an agreement with Chiesi Group to reacquire the rights to codevelop and commercialize its hemophilia B gene therapy in Europe and other select territories and to terminate their codevelopment and license agreement. "Chiesi's decision was driven by recent changes in our strategic priorities," stated Ugo Di Francesco, chief executive officer of Chiesi (Figure 5.25).

5.4.2.2   Zalmoxis

- Discovery and pre-clinical stage
  - *Spin-off from academia*: MolMed was founded in 1996 as an academic spin-off of the San Raffaele Scientific Institute.
- Market authorization
  - *Market authorization in the EU*: Zalmoxis was approved by the EMA as an adjunctive treatment in haploidentical hematopoietic stem

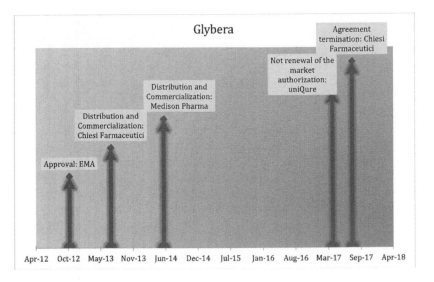

FIGURE 5.25 Partnership agreements for Glybera.

cell transplantation (HSCT) of adult patients with high-risk hematological malignancies in August 2016.

- Post-market authorization
  - ***Manufacturing contract***: MolMed, a medical biotechnology com-
    pany focusing on research, development, manufacturing, and clini-
    cal validation of innovative therapies to treat cancer, and Miltenyi
    Biotec announced in February 2018 that the EMA had approved
    the utilization of the CliniMACS Prodigy, a closed environment,
    automated cell production device developed by Miltenyi Biotec,
    in the GMP manufacturing of Zalmoxis. With the CliniMACS
    Prodigy instrument in its GMP facility, MolMed will be able to
    optimize the efficiency of the Zalmoxis manufacturing cycle in
    terms of doses obtained from a single production cycle and to
    reduce the number of open handling steps, ensuring a reproducibly
    high-quality standard.
- Market withdrawal
  - On 9 October 2019, the European Commission withdrew the mar-
    keting authorization for Zalmoxis (nalotimagene carmaleucel) in the
    EU. The withdrawal was at the request of the marketing authorization
    holder, MolMed S.p.A, which notified the European Commission of
    its decision to permanently discontinue the marketing of the prod-
    uct for commercial reasons. This company's decision considers the
    overall results of the interim analysis voluntarily carried out by the
    company as part of the review of the product development plan, as
    well as the interactions with EMA in recent months. The company

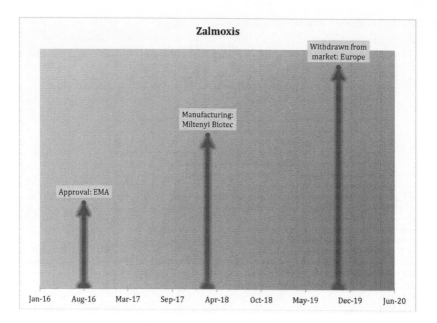

FIGURE 5.26   Partnership agreements for Zalmoxis.

decided to direct its investments and resources, until now destined to Zalmoxis, to activities that better contribute to its business objectives (Figure 5.26).

5.4.2.3   ChondroCelect

- Discovery and pre-clinical stage
  - *Spin-off from academia*: TiGenix was founded in February 2000 by Professor Dr. Frank P. Luyten and Gil Beyen as a spin-off from the Katholieke Universiteit Leuven and the Universiteit Gent.
- Market authorization
  - *Market authorization:* ChondroCelect received market authorization from the EMA for the repair of single symptomatic cartilage defects of the femoral condyle of the knee in October 2009.
- Post-market authorization
  - *Manufacturing contract*: TiGenix completed the sale of its Dutch production facility to PharmaCell BV for a total consideration of €5.75 million in June 2014. PharmaCell acquired the shares of TiGenix's wholly owned subsidiary, TiGenix BV, which holds the Dutch manufacturing facility. ChondroCelect will continue to be manufactured at the facility as before under a long-term CMO agreement with PharmaCell.
  - *Distribution and commercialization*: TiGenix has licensed the marketing and distribution of ChondroCelect to the international

specialty healthcare company dedicated to rare diseases, Swedish Orphan Biovitrum (Sobi), in April 2014. Sobi will continue to market and distribute the product where it is currently available and has also acquired the exclusive rights to expand the product's availability to patients in multiple additional territories, including the rest of the EU, Norway, Switzerland, Turkey, and Russia, plus the countries of the Middle East and North Africa.

- Market withdrawal
  - On 29 July 2016, the European Commission withdrew the marketing authorization for ChondroCelect in the EU, which was effective 30 November 2016. The withdrawal was at the request of the marketing authorization holder, TiGenix NV, which notified the European Commission of its decision to permanently discontinue the marketing of the product for commercial reasons. Due to the regulatory environment around autologous chondrocyte-based cell therapy products in Europe leading to a difficult competitive landscape for ChondroCelect, together with the lack of reimbursement in key European countries, TiGenix was prompted to initiate the withdrawal process of the marketing authorization for ChondroCelect for commercial reasons. Consequently, TiGenix came to an agreement with Sobi for the early termination of their existing commercial relationship and terminated its manufacturing agreement with PharmaCell. This decision was in line with TiGenix's strategy to concentrate its

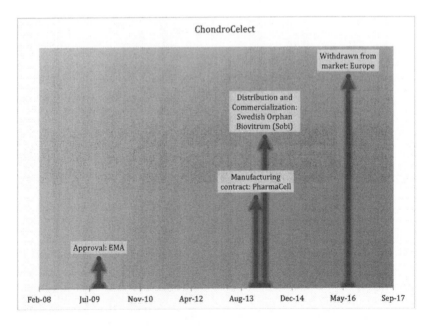

FIGURE 5.27 Partnership agreements for ChondroCelect.

resources and capabilities on its allogeneic stem cell platforms, its upcoming Cx601 phase 3 US trial, and its other clinical stage assets (Figure 5.27).

5.4.3 SUMMARY

All the RMs that are still on the market have collaborations from multidisciplinary organizations: academia, biotechnology companies, service companies, and most important, big pharma. One exception is Zynteglo from Bluebird Bio—this is the only example that no Big pharma was engaged from the research stage to market approval. However, the market launch of Zynteglo in Europe was not advancing as smoothly as expected, manufacturing concerns delayed access in 2019, then the COVID-19 pandemic led to second delays of the market launch, which was expected in half of 2020. However, ultimately, Bluebird Bio decided to withdraw the Zynteglo from the EU market due to commercial reasons. While in the United States, upon BLA submission, the FDA demanded additional information on the manufacturing process the company will use as it transitions the product, LentiGlobin, from clinical trials to commercial production. Thus, the resubmission of MA applications will not be expected before mid-2021. Although this delay is also contributed by the COVID-19 pandemic, the lack of Big pharma involvement could not be excluded.

However, when examining the RMs that have already been withdrawn from the market, it is quite illustrative that one single biotechnology company may be less experienced and less well-prepared to navigate the substantial challenges that exist for RMs.

5.5 CONCLUSION

Partnerships for RMs represent a hybrid model: a collaboration in R&D for assets in the early stage of development, licensing when the proof of concept is validated, manufacturing and logistics services for clinical trials and commercial use, and distribution agreements to expand foreign markets. Collaborations between multiple highly specialized, small players in the RM field with complementary expertise provide fertile ground for boosting the successful translation of scientific discoveries to RM products.

Looking at the key agreement-makers of RMs, we see that biotechnology companies and academia play critical roles in the early stage of development until licensing agreements are established; however, they are also involved in the subsequent stages to provide technological support and expertise; major pharmaceutical companies are bigger players in late-stage clinical trials, commercialization, and distribution, which are achieved either through licensing agreements or outright acquisition of biotechnology companies with late-stage assets; public funding from governments remains paramount for start-up biotechnology and academia to explore the potential of their discoveries, but could provide less leverage for companies that have stable and sustainable pipelines; and service companies provide support throughout the entire development and commercialization journey, considering the unsolved challenges in the manufacturing and logistic processes.

Manufacturing advantages, such as a robust cell-processing technology platform or a novel gene delivery vector, are still the key differentiators to stay competitive in this fierce market. In addition, the potential for allogenic RMs has been increasingly noticed because they bring with them the possibility of scalability and automation. The result may be a reduction in manufacturing costs and potentially a better cost-effectiveness outcome that, in turn, would increase the likelihood of being reimbursed. Undoubtedly, an appropriate disclosure of the proprietary manufacturing methods, as well as a unique discovery platform, will also be important drivers for biotechnology companies to attract either investments or the attention of the major pharmaceutical companies. However, partnering with small players will become more and more complex if they are increasingly acquired by larger players (e.g., major pharmaceutical companies and larger CMOs).

With the expansion of concepts being proven and advancements in technology, RMs may challenge several current "blockbuster drugs" held by the major pharmaceutical companies. In addition, some of these drugs will face patent expiry in the next few years. In order to stay competitive, the major pharmaceutical companies will need to engage increasingly earlier and possibly make in-house efforts to build up highly systematic processes that may even incorporate discovery steps. The RM field is highly dynamic, and the partnerships will continue to grow in number and value, especially as competition becomes increasingly intense in the years ahead. Biopharmaceutical companies that are positioned to take advantages of partnerships will be competitive to deliver their RMs in a fast and efficient manner.

6 Business Models and Commercial Considerations for Gene and Cell Therapies

6.1 FORECAST OF THE CELL AND GENE THERAPY MARKET

According to the report by Emergen Research, the global Cell and Gene therapy (CGT) market was valued at USD 3.21 billion in 2021 and is projected to reach USD 33.40 billion by 2030, growing at a compound annual growth rate (CAGR) of 29.7% from 2022 to 2030 (219). The cell therapy segment was valued at USD 2,237.6 million in 2021 and is expected to be valued at USD 22,631.4 million by 2030 (219), and the gene therapy segment was valued at USD 969.8 million in 2021 and is expected to be valued at USD 10,772.0 million by 2030 (219) (Figure 6.1).

CGT is expected to transform the disease treatment paradigm by treating illness or strengthening an individual's body's resistance to illness by replacing a damaged gene or adding a new gene. Cell and gene therapies have great potential to target both active and chronic conditions, including cancer, cystic fibrosis, heart disease, diabetes, hemophilia, and AIDS. CGTs for cancer currently dominate the market, and the dominance will become more significant in 2030's forecast (Figure 6.2).

Developers from North America lead the global research and development of CGTs, which represented more than half of the market share (Figure 6.3).

Whereas CGTs account for just 1% of launched products in major markets, they represented 12% of the industry's clinical pipeline and 16% of the preclinical pipeline in 2020 (220). Quinn et al. estimated that 350,000 patients will be treated in the United States with 30 to 60 CGT products by 2030 (221). About half the new treatments are expected to be in B-cell (CD-19) lymphomas and leukemias.

In 2022, there were more than 5,000 unique therapeutic products in development and more than 2,260 clinical trials ongoing worldwide. Although the majority of products are in the early phase 1 or phase 1/2 stage, more than 100 products are in the late phase 3 stage (Figure 6.4).

Based on the McKinsey report using data from PitchBook, venture capital biotech funding peaked in the first quarter of 2021 and has slightly declined since. CGTs have dominated the biotech investment globally. Medical biotech companies that develop CGT platforms attracted two-thirds of the $52 billion of venture capital (VC) investment from 2019 to 2021 (Figure 6.5) (222).

DOI: 10.1201/9781003366676-6

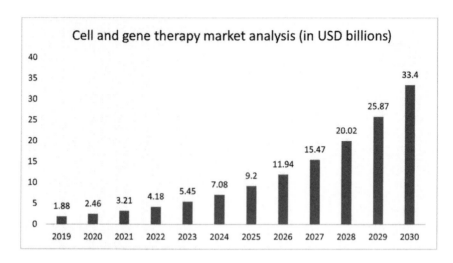

FIGURE 6.1   The market estimate forecast for cell and gene therapies.

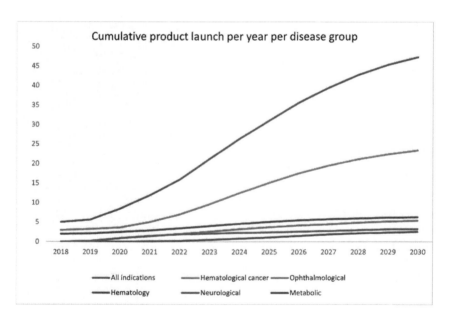

FIGURE 6.2   Cell and gene therapy market estimate forecast by indication.

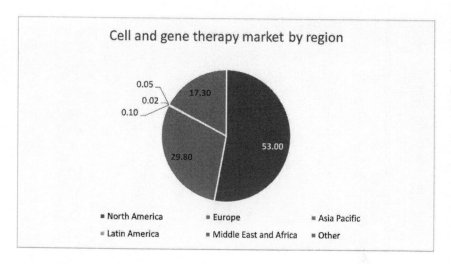

FIGURE 6.3 The distribution of the cell and gene therapy developers by location.

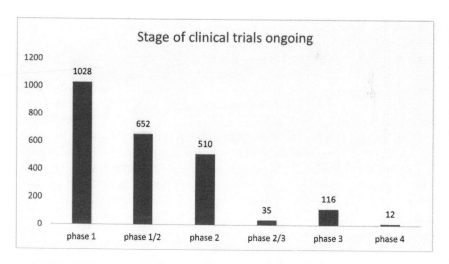

FIGURE 6.4 Stage of clinical trials for cell and gene therapies ongoing collected from American Society of Cell and Gene Therapy as of September 2022.

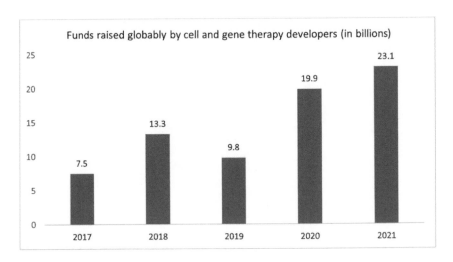

FIGURE 6.5   Funds raised globally cell and gene therapy developers (in USD billions).

6.2   INCREASING INVESTMENT WILL BRING FAST-GROWING REVENUE

Investment in GCT can dramatically increase the market valuation of a company. For example, when Novartis acquired AveXis for 8.7 billion, their stock market increased by about $9 billion. Novartis is one of the largest pharmaceutical companies that are most engaged in gene therapy. They have launched Kymriah and Zolgensma in multiple countries and still are investing in several products. They have created Novartis Gene Therapy as a more or less independent entity, which is dedicated to developing and commercializing the gene therapies for rare and life-threatening neurological diseases (223).

Also, encouraging results from clinical trials for CGTs will boost the market value of the developers. For example, Biomarin gained a $1 billion market cap after positive results were released for their gene therapy for hemophilia A (224). Further, uniQure's share gained 20% in value with initial positive results in a Huntington disease gene therapy trial (225). Similarly, LogicBio's share gained 60% with the news that the company's phase 1/2 gene therapy trial would be resumed (226).

The current CGT landscape is often compared to the history of the monoclonal antibody therapy (mAb). Whereas the first products in these two categories were launched two decades apart, their sales projections seem to follow a similar pattern: annual sales of about $5 billion and between 10 and 20 assets in the first several years (220).

One way to contextualize the substantial growth in investment in cell and gene technologies is to look at forecasted revenue growth for the therapies that are in development. Based on worldwide sales forecasts from Evaluate Pharma, it is expected that conventional drug sales will grow at a CAGR of 6% from 2021 to 2026 and biologic sales, excluding CGTs, are forecast to grow from $415 billion to $541 billion, a CAGR of 5%. CGTs, in comparison, are expected to grow from $4 billion a year in sales to over $45 billion over that same period, a significantly higher CAGR of 63%.

6.2.1 Cell and Gene Therapy Market Forecast for the United States

As the largest market, U.S. CGTs market size is expected to reach $14,887.0 million in 2030, indicating a CAGR of 28.4% for revenue of CGTs (Table 6.1 and Figure 6.6).

6.2.2 Cell and Gene Therapy Market Forecast for Europe

The CGT market size in Europe is expected to reach $10,956.3 million in 2030, indicating a CAGR of 31.1% in terms of revenue (Table 6.2 and Figure 6.7).

6.2.3 Cell and Gene Therapy Market Forecast for China

The CGT market size in China is expected to reach $1,519.7 million in 2030, indicating a CAGR of 33.6% in terms of revenue (Table 6.3 and Figure 6.8).

TABLE 6.1

US Cell and Gene Therapy Revenue Estimates and Forecast

Market Size	2019	2020	2021	2022	2025	2028	2030	CAGR%
Revenue ($ million)	928.8	1,203.0	1,556.2	2,010.4	4,301.2	9,095.4	14,887.0	28.4%

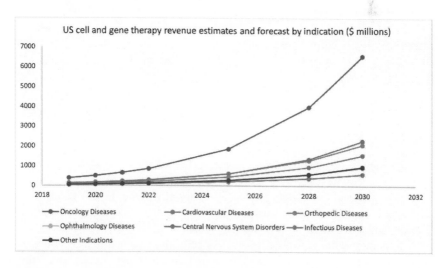

FIGURE 6.6 US cell and gene therapy revenue estimates and forecast by indication (USD millions).

TABLE 6.2

Cell and Gene Therapy Revenue Estimates and Forecast in Europe

Market Size	2019	2020	2021	2022	2025	2028	2030	CAGR% (2022–2030)
Revenue ($ millions)	546.8	723.0	954.6	1,258.7	2,862.4	6,430.5	10,956.3	31.1%

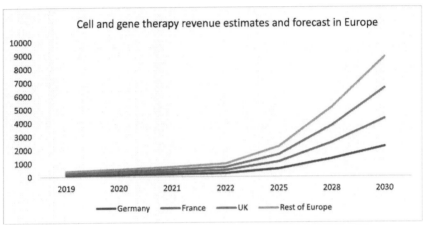

FIGURE 6.7   European cell and gene therapy revenue estimates and forecast ($ millions).

TABLE 6.3

Cell and Gene Therapy Revenue Estimates and Forecast in China

Market Size	2019	2020	2021	2022	2025	2028	2030	CAGR% (2022–2030)
Revenue ($ millions)	61.1	82.5	111.3	149.9	362.0	860.5	1,519.7	33.6%

6.3   MORE CHALLENGES IN MANUFACTURING AND DELIVERY

6.3.1   CHALLENGES IN MANUFACTURING

Gene therapy is the use of genetic material in the treatment or prevention of disease. The genetic material transferred into the patient body changes how a single protein or group of proteins is produced by the cells. Gene therapy can be used to reduce levels of a disease-causing version of a protein, increase production of disease-fighting proteins, or produce new proteins (227).

Cell therapy is the transfer of intact, live cells into a patient to treat a disease. If the cells originate from the patient, they are called autologous cells, and if they come from a donor, they are called allogeneic cells. The cells are also classified by their potential to transform into different cell types: pluripotent cells transform

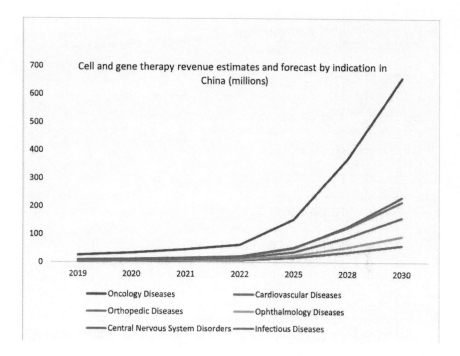

FIGURE 6.8   China cell and gene therapy revenue estimates and forecast by indication ($ millions).

into any cell type in the body, and multipotent cells can transform into other cell types, but their repertoire is more limited (227).

CAR-T is a specific kind of cell therapy. The term stands for chimeric antigen receptor therapy (CAR-T). This method modifies the patient's own immune cells (T cells) to express a receptor on their surface that recognizes antigens on the surface of malignant cells. Once the receptor binds to a tumor antigen, the T cell is stimulated to destroy the malignant cells (227). Nearly half of the CGTs undergoing clinical testing are cancer treatments. CAR-T treatment modifies the DNA of a patient's immune cells such that they react to cancer cells. The immune cells are then reintroduced, allowing the patient's immune system to combat the malignancy (219) (Figure 6.9).

The newest wave of treatments represents "CGT 3.0," building on the dendritic cell and CAR-T cell breakthroughs that have already received regulatory approval over the last 12 years. Many of the newest approaches rely on manipulating a broader range of cells—such as allogeneic cells from donors or starting material from solid tumors—that are often even more complex to collect, manage, and dose than those used in CAR-T treatments (228).

With many CGTs, manufacturers are producing a highly complex product that is delivered on a personalized basis to each patient. CGT manufacturing processes take the inherent variability of the living product into account. For example,

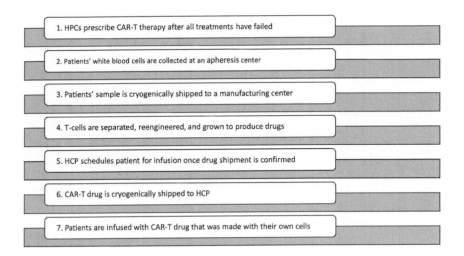

1. HPCs prescribe CAR-T therapy after all treatments have failed

2. Patients' white blood cells are collected at an apheresis center

3. Patients' sample is cryogenically shipped to a manufacturing center

4. T-cells are separated, reengineered, and grown to produce drugs

5. HCP schedules patient for infusion once drug shipment is confirmed

6. CAR-T drug is cryogenically shipped to HCP

7. Patients are infused with CAR-T drug that was made with their own cells

FIGURE 6.9   Complex manufacturing process of CAR-T cell therapies.

depending on the technology in question, manufacturers must consider additional manufacturing challenges, such as genetic stability and tumorigenicity, immunogenicity, and off-target effects, among others. Moreover, for those CGTs that are personalized treatments, the investment must be made many times over for each individual asset (229).

Reportedly, the reliability of the external cell manufacture can be as low as 33%. That means a doctor has to tell two out of every three patients, after they had undergone apheresis, that there was no product and therefore no treatment.

6.3.2 Challenges in Patient Selection

Regarding delivery and administration, CGTs require careful patient selection, as well as multiple detailed preparation and planning steps to ensure optimal treatment. For example, stem cell–based therapies for neurological disorders, such as Parkinson disease, are very complex procedures involving not only a carefully produced and characterized product but also tailor-made delivery devices and highly specialized neurosurgeons capable of administering the treatment. Administration of the final product into a patient's brain may require sophisticated live imaging, as well as customized delivery devices that enable precise injection of therapeutic material. Finally, post-administration, patients need to be monitored for years to allow both the detection of the clinical benefit and any potential safety events. (229) As for autologous cell therapies, there is a complex process that wraps around the delivery of the cell therapy itself and often requires biopharma companies to develop end-to-end services and help build the infrastructure and capabilities needed to deliver therapy these to patients in a safe and effective manner (230).

6.3.3 CHALLENGES IN LOGISTICS AND PRODUCT DELIVERY

By giving their cells and receiving them back post-processing, the patient become part of the manufacturing process. Then cells are sent back to the physician, who will administer them to the patients. In addition to administering CGTs, moving the therapy itself to the treatment site can be complicated. Companies rely on third-party logistics companies to move specimens to and from treatment sites. The number of service providers and the service footprint would have to increase drastically once CGTs are produced at scale (231). This is done centrally and requires the carrier to organize the transport, which was highly disturbed during the COVID-19 pandemic. Delivery is patient specific and therefore requires flexible, adaptable manufacturing, storage, and finished product production for each specific patient.

6.3.4 CHALLENGES IN MANUFACTURING ALLOGENEIC THERAPIES

Considering that the manufacturing process for some products, such as Kymriah, may be long, several patients have received traditional chemotherapy while waiting for CAR-T cell therapy. Therefore, shortening the manufacturing process is a challenge—developing patients' allogeneic CAR-T cells is one promising solution to narrow this time gap. Since allogeneic cells can be used to treat multiple patients and can be manufactured at an industrial scale, they can be offered off-the-shelf, significantly reducing the cost of treatment per patient. For these reasons, the number of new pre-clinical/clinical trials involving allogeneic therapies has been growing at an average of 12.5% each year since 2017. However, despite considerable progress being made in allogeneic cell therapy development, drug developers are struggling with complex manufacturing bottlenecks at various stages of development. These obstacles are limiting the efficacy, safety, and scalability of allogeneic cell products.

It's important to distinguish the various types of CGTs, because each of them poses unique challenges to manufacturers. The manufacturing process for allogeneic cell therapy is different from autologous therapy in several aspects (Table 6.4). Unlike autologous therapies that acquire the starting material from the patient, allogeneic therapies acquire cells from healthy donors. This allows the quality of starting material to be optimized via donor profiling (231). Moreover, allogeneic therapy faces its own issues in product tracing and long-term cryopreservation. A single batch of allogeneic cells is used for multiple patients and sometimes at different clinical sites. Therefore, robust monitoring of various parameters across the end-to-end workflow is essential to ensure effective cell production and identify potential logistical problems. One of the considerations is standardizing information from at least three sources from the manufacturing unit through to the logistic provider to the clinical sites. This can be mitigated via implementing a fully digitalized batch manufacturing record with supply chain integration (232). Scalability when transitioning from the research phase to the clinical production phase is the top challenge for allogeneic cell products. More

TABLE 6.4

Types of Cell and Gene Therapies

Cell Therapy	Gene Therapy
Allogeneic	• Non-viral Vector
• Mesenchymal Stem Cells	• Oligonucleotides
• T-cells	• Viral Vector
• Hematopoietic Stem Cells	○ Retroviral Vectors
• Natural Killer Cells	○ Adeno-associated Virus Vectors
• Induced Pluripotent Stem Cells	○ Other Viral Vectors
• Other Allogeneic Cells	
Autologous	
• Mesenchymal Stem Cells	
• T-cells	
• Hematopoietic Stem Cells	
• Natural Killer Cells	

specifically, immune cells can only be expanded for defined times before reaching senescence, or the exhaustion state, and this biological constraint limits the total number of cells per batch produced from each donor collection. After production, there are also considerations that need to be taken in regard to cryopreservation, distribution, and the traceability of the end products (232).

6.4 WHY CELL AND GENE THERAPIES REQUIRE A UNIQUE BUSINESS MODEL

6.4.1 COLLABORATION TO NAVIGATE MANUFACTURING CHALLENGES

CGTs are challenging, risky, and expensive to manufacture due to the lack of commercial scale of the technologies involved (231). The immature manufacturing sector for gene therapy will force companies to make some bets to circumvent current capacity constraints and meet demanding timelines. Biopharma companies' traditional business and operating models are often not optimal for bringing to market highly customized products like CGTs, which typically require precise material sourcing, tracking, and delivery. For example, the CAR-T process involves several external third parties, such as the center doing the cell collection, and the processing of the cells is done in the same center by another team, based on company procedure. As a result, despite the strong interest in and the promise of CGTs, companies have not yet defined—much less built—infrastructure that supports their manufacture and delivery at scale (231).

6.4.2 LIMITED ROLE OF REGIONAL AFFILIATES

In CGT, the function of regional affiliates, which traditionally play a fundamental role in the marketing, distribution, and payment of drugs sold in specific countries, is uncertain. Interviewees said that given the customized nature of cell therapies, the small volumes, and the short turnaround times, regional affiliates are unlikely to play a role in distribution. Instead, direct-to-patient distribution will likely be centralized through manufacturing facilities or regional hubs (231). Despite the fact that regional affiliates could possibly continue to play a role in offering specific services to local markets—for example, in certifying and engaging treatment sites or tracking patient outcomes after the delivery of therapy—even these services may not make sense at a regional level because of the small volumes of therapies being delivered (231). This is a disruption to the current business model.

6.4.3 DISRUPTIONS TO THE TRADITIONAL PROCESS OF ADMINISTRATION AND MONITORING

Companies that are willing to differentiate their product from competitors in the same disease area may need to consider that their success may depend on how easy the technology is to employ by the healthcare staff. DePinto et al. pointed out that there are a number of issues with CGTs that should be addressed to improve their willingness to prescribe CGTs:

- Workflows and operational systems are not standardized for critical steps across the CGT patient and product journeys, introducing extra work, complexity, and risk.
- Healthcare providers (HCPs) are losing excessive amounts of uncompensated time to "back office" work, including duplicative IT audits and risk assessments, one-off system trainings for each individual CGT product, and repetitive, high-risk manual data entry that all lead to delays in offering therapies.
- Clinical staff is overwhelmed by the need to prioritize processes and training over time with patients, resulting in low morale and attrition.
- IT and cybersecurity vulnerabilities are arising from the proliferation of too many individual manufacturer portals and digital systems (228).
- CGT treatment sites must not only be audited and undergo Foundation for the Accreditation of Cellular Therapy accreditation but must also train personnel on drug profiles, procedures for sample collection and shipment, and portals to interact with the manufacturer. This is a time-consuming process for the clinical staff (231).

The complexity in the administration of CGTs also adds extra burdens on patients. First, the FDA requires clinics to follow up with patients who receive CGTs for 15 years after the administration—much longer than with regular biologics. Second, patients may travel long distances to receive treatment in specialized centers. Third, additional complexities can arise when patients change providers or health insurers (231). Further, patients can be admitted and treated as inpatients or as outpatients. In the latter case they must have suitable housing nearby the clinic.

6.4.4   The Time of Approval Is Critical to Secure Market Success

Business sustainability can also be viewed from the product/service perspective, i.e., how well the product can sustain competition, the availability of alternative treatment options, and so on (233). Currently, over two-thirds of the products being developed are for rare diseases and the rest are for non-rare indications. For gene therapies that are focused on ultra-rare/rare disorders, there is a question of commercial viability of the business model given the small number of incident patients, some of whom may be treated in the trial itself. First to market is an important topic for rare conditions, as the first treatment will treat the prevalent cases while follow-up products will share the incident cases. As rare diseases usually have a low incidence, the size of the target population drops very significantly once the prevalent patients have been treated with regenerative therapies (234).

For CGTs targeting cancer, the prevalent and incident pool of patients is much larger than for rare diseases, but the need to customize, iterate, and redose patients based on their tumor profiles presents a very different approach to research and development. In other words, gene therapies for cancer face intense competition from existing therapies available at a lower cost. For instance, Kymriah (tisagenlecleucel) has several competitors, such as blinatumomab (Blincyto, Amgen), inotuzumab ozogamicin (Besponsa, Pfizer), and several other products that are available at significantly lower cost (Blincyto at US\$ 178,000 in 2014, Besponsa at US\$ 168,300 in 2017, and Kymriah at US\$ 475,000 in 2017).

Knowledge, know-how, and science are growing fast and being disseminated widely, which will lead to new generation of such products to emerge rapidly, making first-generation products obsolete. Therefore, the lifecycle of a gene therapy can be significantly shorter than that of a regular, long-term treatment, which has to be reflected when planning pipelines (235). In fact, every step of the clinical regulatory development plan for a gene therapy must be executed with shorter deadlines than for regular medicines (235).

6.4.5   Different Business Strategies for Cell and Gene Therapies

Large biopharma companies may need to employ unique decision-making and resource allocation approaches for CGTs. It's important to ensure that CGT units in companies are appropriately sourced, even if they operate as part of a larger division. To offset this impact of potential returns on funding decisions, large biopharma companies should consider longer-term returns on advancing a portfolio of products, rather than individual ones (231).

Success rates for CGT products are higher than those for small-molecule products, probably because CGT tends to target specific disease drivers rather than the broad targets (with potential for off-target effects) of small-molecule therapy (Table 6.5). Between 2008 and 2018, the research and development (R&D) success rate from phase 1 to launch for small-molecule products was 8.2%; for CGT products, it was 11% (220). Large pharma companies will need to assess if higher revenues justify the higher risks and costs or whether lower risks and costs at

TABLE 6.5

The Comparison of Risks Related to the Development of GSTs and Small Molecules

	Discovery	Pre- and Clinical Trials	Approval	Manufacture and Distribution	MAPR	Patient Follow-up
Costs	Not sure, as usually licensed from academic start-ups	May be smaller due to shorter duration and small population	Lower/same?	Very high	May ask long-term RW data, outcome-based models	High costs (15 years of follow-up ordered by FDA), new IT infrastructure, HCP should be paid for that?
Risks	Slightly more success rate than in small molecules	Last-line treatments + rare conditions have few patients—slow pace	Increased risk due to inexperience and vague regulatory requirements	Very high, living material, identity and custody issues	Refusal due to high price, novel payment models may take time to agree on, price erosion due to competition, genes cannot be patented	HCP not compliant, administrative burden, irrelevant/poor data

certain stages of the R&D process could compensate for higher costs and risks at other stages (Table 6.5).

In the meantime, biotech companies remain leaders in CGT innovation. Only 2 of the top 20 biopharmaceutical companies have CGT assets, making up more than 20% of their pipelines. As of February 2020, only a small percentage of launched CGT assets either originated from or are owned by the top 20 biopharmaceutical companies—in both cases, only 15% of these launched assets, indicating how much opportunity there is for such companies to increase their exposure to CGT assets (220).

6.5 PARTNERSHIPS TO DRIVE THE CGT LANDSCAPE PROGRESSION

6.5.1 THE ADVANTAGE OF PARTNERSHIPS

Companies need to invest in delivery platforms that have the potential to supply a stream of assets that require only minimal modification. They cannot rely on internal product development alone; they will need to draw on work underway in academia, biotech, and other areas. Notably, 58 of 120 registered gene therapy phase I clinical trials have an academic sponsor (234). More often than not, the original research behind new platforms is conducted by academics (who set up their own biotech companies) and investors (whose models include company origination because of the potential financial gains and the concentrated technical risk that platform investments carry). VC firms are more comfortable than established biopharma with such risks (220).

Network partnerships are important for aspects of the CGT value chain, which may be too complex or costly for CGT companies to develop competencies in, including specimen collection and delivery logistics (231). Partnerships played important roles in facilitating and streamlining the market access of CGTs for several reasons:

- Start-ups are often first entering this field, and they are insufficiently funded so partnerships will lower the need for initial investments.
- The development may require several competencies such as viral vector production and processing cells that are complex to develop from scratch and require a chain of expertise that would be expensive and underused if managed in-house by large pharma companies.
- The development of such facilities by one company from the green field will be underused for long time, making them an expensive investment with lack of return especially at the time of development.
- Partnerships also bring cash, know-how, etc., to young start-ups.

Partnerships structured to give a biopharma company access to a single asset are the simplest way to enter the CGT market and are often chosen by companies that have a strong focus on certain indications and believe that their competitive advantage lies in owning multiple therapies across modalities in that space. A single-asset partnership also minimizes the investment required. However, this kind of partnership may leave a biopharma having to introduce a new operating model for a single asset. Partnerships structured to give a biopharma access to all assets from a platform in certain therapeutic areas to help companies with a strong strategic focus on a given therapeutic area strengthen their portfolios and build more expertise in that area. In addition, more assets in a new modality mean more opportunity to build the relevant development and commercial expertise (220).

Focusing in this way could enable companies to use a plug-and-play approach for their pipeline. Such platforms allow for better management of costs, consistency in processes, and timeline efficiency. For example, Pfizer has built out an adeno-associated virus (AAV) platform with a combination of acquisitions (Bamboo and Vivet) and strategic licensing partnerships (with Spark Therapeutics and Sangamo), while making substantial internal investments (more than $500 million in manufacturing). Sarepta has built its pipeline by licensing a series of assets from academic institutions and smaller biotech firms (including Myonexus,

Lysogene, and Lacerta), while developing an external network for manufacturing with companies such as Brammer, Paragon, and Aldevron (234).

Nowadays, we observe a trend toward market consolidation with large pharma companies acquiring assets through company acquisitions or partnerships. Large players in the pharma business are now multiplying the number of assets they develop or commercialize. This might make the concentration of expertise more common and the need for partnerships less and less frequent. However, during at least a decade, one can expect that both models, that is integration in large pharma companies and partnerships with start-ups, will co-exist.

6.5.2   KEY CELL AND GENE THERAPY PARTICIPANTS

- **Roche.** The company has developed to become one of the world's largest biotech firms, as well as a leading provider of in vitro diagnostics and a global developer of transformational new solutions in significant disease areas. The business contributes to the advancement of the production and process development of CGTs, from cell isolation to quality control release testing.
- **Novartis AG (Novartis).** Novartis Institutes for BioMedical Research (NIBR), a division of Novartis, performs research on a variety of diseases. Novartis gene therapy is devoted to developing gene therapies with the potential to positively impact the lives of the patients and families devastated by rare and life-threatening neurological genetic diseases.
- **Intellia Therapeutics** is a well-known biotechnology company which specializes in the development of clustered regularly interspaced short palindromic repeats (CRISPR) genome editing technologies and is rapidly moving experimental therapies into the clinic. The company was formed to lead the biotechnology industry in one of the most promising new therapeutic areas, gene editing and repair using CRISPR-Cas9 technology.
- **Sarepta Therapeutics.** Sarepta discovers and develops novel RNA-targeted medicines for the treatment of rare diseases. The company creates pipeline products in gene therapy, RNA, and gene editing using its multiplatform Precision Genetic Medicine Engine. The platform is based on the company's ground-breaking work with phosphorodiamidate morpholino oligomer (PMO) chemistries. SRP-9001, SRP-5051, SRP-9003, and SRP-5045 are in development for the treatment of Duchenne muscular dystrophy (DMD), limb-girdle muscular dystrophies (LGDMs), and other neuromuscular and central nervous system disorders.
- **Regenxbio** is working on gene therapy product candidates for the treatment of retinal, metabolic, and neurodegenerative diseases. The products are all made with AAV viral vectors from the company's proprietary gene delivery platform, known as the NAV Technology Platform. Aside from internal product development, the company selectively licenses its NAV vectors to other leading biotechnology companies.

Players in the CGT market employed various strategies like partnerships and agreements, product approval, investment, and acquisition over the study period of 2019–2022 in order to increase their market share by reaching out to a newer portion of the potential consumer base, as well as holding the current consumer base through various tactics. Of all the strategies employed during the review period, acquisition by the major companies is one of the leading strategies followed by partnerships and acquisitions (Table 6.6).

TABLE 6.6
Strategy Initiatives in Cell and Gene Therapies for Key Participants 2019–2022

Company	Date	Type	Details
Roche	July 2022	Research	Roche and its partner Sarepta Therapeutics revealed positive functional data from a number of trials of their gene therapy for Duchenne muscular dystrophy (DMD).
Novartis	June 2022	Product approval	Novartis announced that the European Commission (EC) has approved Tabrecta (capmatinib) as a monotherapy for the treatment of adults with advanced non–small cell lung cancer (NSCLC) that has mutations that cause exon 14 (METex14) skipping and who need systemic therapy after receiving prior immunotherapy and/or platinum-based chemotherapy.
Novartis	June 2022	Announcement	Novartis announced the publication of Zolgensma data in *Nature Medicine* demonstrating age-appropriate benchmarks when treating kids with SMA pre-symptomatically. Pre-symptomatically treated children with two or three copies of the *SMN2* gene achieved age-appropriate milestones such as sitting, standing, and walking.
Regenxbio	June 2022	Investment	Regenxbio announced a USD 65 million investment in the construction of a new gene therapy manufacturing facility in Montgomery County, Maryland.
Sarepta	June 2022	Announcement	Sarepta announced that the FDA has placed a clinical hold on SRP-5051 (vesleteplirsen), the company's next-generation peptide-conjugated phosphorodiamidate morpholino oligomer (PPMO), for the treatment of patients with Duchenne muscular dystrophy who are amenable to exon 51 skipping.

TABLE 6.6 *(Continued)*
Strategy Initiatives in Cell and Gene Therapies for Key Participants 2019–2022

Company	Date	Type	Details
Sarepta	May 2022	Announcement	Sarepta announced the launch of The Limb-Girdle Muscular Dystrophy (LGMD) Grant Award Program. Its creation was intended to speed up the LGMD diagnostic process and increase participation in current genetic testing programs. Patient advocacy groups and non-governmental organizations will get funding from the program to support their suggested activities.
Intellia	February 2022	Acquisition	Intellia announced the acquisition of Rewrite Therapeutics, a biotechnology company based in Berkeley, California, that focuses on developing novel deoxyribonucleic acid (DNA) writing technologies. The acquisition adds a platform that is highly complementary to Intellia's existing CRISPR or Cas9 and base editing technologies, further expanding the company's industry-leading genome editing toolbox.
Roche	September 2021	Partnership	Roche and US-based biotechnology company Shape Therapeutics entered into multitarget strategic cooperation and license agreement to explore gene therapy for specific targets in Parkinson disease, Alzheimer disease, and rare disease areas. Shape is now eligible for more than USD 3 billion in initial and prospective payments from Roche for meeting research, regulatory, and sales objectives.
Regenxbio	September 2021	Partnership	Regenxbio announced a strategic alliance with AbbVie, which is a leading research-based biopharmaceutical company headquartered in North Chicago, Illinois. The collaboration aims to develop and commercialize RGX-314, which is a potential one-time gene therapy for wet age-related macular degeneration (AMD), diabetic retinopathy (DR), and other chronic retinal diseases. Under the terms of the agreement, Regenxbio and AbbVie shared the costs of future RGX-314 trials, including the planned second pivotal trial evaluating subretinal delivery for the treatment of wet AMD.

(Continued)

TABLE 6.6 *(Continued)*
Strategy Initiatives in Cell and Gene Therapies for Key Participants
2019–2022

Company	Date	Type	Details
Roche	August 2021	Acquisition	Roche has signed a share purchase agreement to acquire Irish biotech business Inflazome for USD 449 million. The acquisition provides Roche complete control of Inflazome's entire portfolio of clinical and pre-clinical orally accessible small molecule NLRP3 inhibitors. Roche plans to continue developing NLRP3 inhibitors for a wide range of indications with significant unmet medical needs.
Roche	May 2021	Acquisition	Roche announced the acquisition of GenMark Diagnostics for USD 1.8 billion. With GenMark's experience in syndromic testing, Roche hopes to increase the scope of its molecular diagnostics offering with the acquisition. Roche's efforts to manage infectious diseases and antibiotic resistance will be improved by GenMark's ePlex systems.
Roche	November 2020	Agreement	CEVEC Pharmaceuticals has agreed to use Roche's recently announced Elevecta Technology for large-scale manufacture of adeno-associated virus (AAV) vectors for gene therapy applications under an option and license agreement. Under the terms of the agreement CEVEC will provide Roche and Spark Therapeutics, which is a part of the Roche Group, an option for a nonexclusive license to create AAV producer cell lines based on CEVEC's Elevecta Technology.
Novartis	July 2020	Acquisition	Novartis successfully acquired The Medicines Company, adding the investigational cholesterol-lowering therapy inclisiran, which may be a first-in-class treatment. With the acquisition of The Medicines Company and inclisiran, Novartis has the opportunity to write a new beginning in the development of vaccine-like treatments for the world's top cause of mortality and disability.

6.6 NEW PAYMENT MODELS ARE NEEDED FOR SUCH INNOVATIVE AND COSTLY THERAPIES

6.6.1 THE EFFICACY AND SAFETY OF CGTS ARE NOT FULLY UNDERSTOOD

Clinical development programs for CGTs tend to differ from those adopted for previous treatment modalities. Whereas randomized controlled trials have long been considered the gold standard for evidence generation, this approach poses several challenges for the clinical evaluation of many CGTs due to small patient populations currently targeted, the absence of relevant therapeutic alternatives for patients, ethical considerations regarding clinical equipoise vs. standard of care, and potentially the use of endpoints that have yet to be validated in new therapeutic areas but which capture critical benefits conferred by these new treatments (229). Furthermore, as blinding may not always be feasible or not maintained because of the lack of efficacy, patients drop out of the clinical trial, making it de facto a failed trial and very difficult to interpret.

Also, most CGTs have limited long-term efficacy data relative to their long-term value proposition. Consequently, the evidence packages for CGTs tend not to fit within the existing health technology assessment (HTA) frameworks that are used to evaluate a product's value for money. As a result, payers foresee risk in terms of whether a CGT will be as effective and durable as proposed in its value proposition (i.e., whether the proposed value for money will be realized) (229).

6.6.2 DIFFERENT VALUE PROFILES FOR DIFFERENT PATIENT POPULATIONS

The advancement of a technology does not indicate its value to healthcare payers. For example, the first approved therapeutic cancer vaccine, Provenge, failed to achieve commercial success and was subsequently withdrawn from the market, in spite of being a scientific breakthrough (236). On the other hand, a traditional, small molecule drug may attract relatively high prices if it is believed to have great clinical value. Illustratively, the sedative thalidomide, in a superior form called lenalidomide, as a multiple myeloma treatment in 2005 sold for $280,000 for a year of treatment (237). Not only did the drug respond to an unmet clinical need but cancer treatments tend to achieve higher prices because it is a particularly dreaded disease area (237). Similarly, drugs for diseases affecting children represent higher value to payers because their benefits can be accrued over longer periods of time, e.g., a drug approved for a newborn that improves their quality of life (QoL) might produce sustained benefits for over 80 years, but a drug for a 50-year-old person for only 30 years. Similarly, a lifesaving drug will save more years of life if given to a child than to an elderly person.

Defining the target population for a novel CGT is a challenge. So far, CGTs have only proved their potential in rare diseases with well-defined genomic targets and high unmet needs. Going for a rare condition limits the budget impact of such high price therapies, which facilitates pricing and reimbursement negotiations with payers. However, the potential of CGT platforms mush be expanded to diseases with larger prevalence indications such as neurological disorders, diabetes, and cardiovascular illness. This will allow the ability to maintain funding, resources, and attention (238). However, going for a larger indication will result in very tough price negotiations because of the high budget impact that may threaten the sustainability of health insurers. Successful CGT will need to demonstrate (cost-)effectiveness against traditional modalities of treatment and care, ensure superior delivery methods, and cope with scaling up production (238).

Further, severity of the condition, life expectancy at the time of treatment, progressive nature of the disease, vulnerability of the population (babies, for example), lack of availability of alternative therapies, etc., are important dimensions that drive the value recognition by payers and other stakeholders. Early engagement with payers and value assessors to understand how they define value is critical to incorporating the right endpoints into clinical trials (231).

6.6.3   High Prices of Cell and Gene Therapies Raised Payment Barriers

CGTs come at unprecedently high prices, for example, Zolgensma, the gene therapy treatment for spinal muscular atrophy (SMA) commercialized by Novartis, is priced at US$2.125 million, and Zyntenglo, the gene therapy for β-thalassemia commercialized by Bluebird, is priced at $2.8 million (231). At the time of writing, the FDA granted accelerated approval for Skysona, or eli-cel, for the rare neurological disorder cerebral adrenoleukodystrophy (CALD). The company is expected to charge $3 million per treatment with Skysona, higher than Zynteglo's $2.8 million, making it the priciest therapy in the world. Such high prices reflect the high cost and complexity of developing CGTs, as well as the significant value for patients and their families that are in urgent need of lifesaving treatment.

CGTs are often developed for children and have the potential to both improve the QOL and/or extend their life. However, despite benefits being produced over a long time, the drugs are delivered in a single administration (239). Because drugs are typically paid for up-front, this implies that large sums of money need to be spent by payers at once. Therefore, payers may face affordability issues as they need to work with an annual drug budget (229).

In particular, the concern over budget impact is a direct result of recent payer experience with innovative therapies for conditions like hepatitis C. The high level of unmet need when the first of these products was launched in 2013, particularly for populations covered by Medicare and state Medicaid plans, created significant pressure on payer budgets. As more products came to market, increasing competition brought prices down, but the lesson for gene therapy is clear: carefully planning for and managing utilization is critical (240).

High one-time costs would make it challenging for payers to underwrite the risk of full payment for the entire range of gene therapies coming to market simultaneously. In addition, a fixed, up-front pricing leaves the payer with all the risk of the therapy not working, given that the long-term efficacy of gene therapy, as well as the risk of toxicity and other harmful effects to patients, are not known with certainty at the time of first regulatory approval (240).

Presently, commercial uptake is highly uncertain due in part to the already-described mismatch between existing HTA frameworks/drug reimbursement models and the clinical trial designs typically employed for CGTs (i.e., comparatively small, sometimes single-arm trials that observe benefit over a shorter period of time than the duration of benefit suggested in their value proposition). For example, the French pricing committee decided that the price of single-administration therapies will be based on the duration of efficacy at the time at launch captured in the clinical trials (241). If this gap leads to delayed access, reduced access, or denial of access, manufacturers will face challenges in the commercial predictability, stability, and overall viability of bringing new treatments to market. Due to the lack of effective strategies, products such as ChondroCelect, Glybera, and Zyntenglo were all removed from the European market due to commercial challenges. Without a clear and viable commercial path, investment in developing future life-altering CGTs will dwindle (229).

Another study found discrepancies among coverage restrictions among various US health plans (242). Also, the European and Canadian HTA bodies recommend reimbursement for fewer therapies than US health plans, reflecting a more science-based approach to decision-making. The scientific approach acknowledges that the long-term durability of patient response for CGT is not yet known (242). In contrast, when coverage is recommended, it is typically consistent with the criteria included in regulatory approvals, with only a few exceptions where HTA bodies applied restrictions beyond the regulatory label. The EU5 and Canadian HTA bodies make market access recommendations reflecting scientific evidence that is appraised according to explicit HTA decision analytic frameworks. Specifically, most HTA agencies continue to rely heavily on evidence from higher-quality studies, such as randomized controlled trials, in their assessments of benefits, risks, and value.

While the EU has unveiled a pharmaceutical-strategy document that recognizes cell and gene therapies as "major milestones" of progress in healthcare, it has also stressed the need to improve the affordability and cost-effectiveness of medicines, as well as health systems' sustainability. The practical challenges are evident in Germany, where Bluebird Bio is withdrawing its β-thalassemia treatment, Zynteglo, because of an inability to agree on pricing and reimbursement (238). This is followed by a broader withdrawal of Zynteglo and Skysona (for the treatment of CALD) from Europe.

The benefits of gene therapy could last years or even a lifetime; however, this poses a challenge for the treatment when it comes to portability. Insurers in countries where the insurance market is fragmented like in the United States, who pay for a one-time treatment may lose patients to another insurer, along with the benefit of having a healthy patient in their risk pool (231). A payer could find itself bearing all the costs but appreciating few of the subsequent benefits when the patient switches to a different insurer. Those benefits would ensue to the new insurer, which would now have a healthier patient in its insurance plan (240).

Also, the reimbursement of many biologic therapies in the United States today occurs under a buy-and-bill model. A treatment facility purchases the therapy for a fixed price, a physician administers it, and the facility then receives reimbursement. The physician receives a separate administration fee. This process can take 30 days or longer, potentially limiting the pool of healthcare providers with enough working capital to assume the reimbursement risk for high-cost gene therapy. In addition, some health plans only cover products in treatment facilities that have negotiated lower reimbursement rates. Given the lower margins, these facilities may be less likely to want to take on reimbursement risks unless they are confident they can attract enough new patients to reduce their costs below reimbursement rates. A smaller provider pool may limit the adoption of gene therapy to nearby patients or those willing to travel but may result in higher-quality, specialized care (240).

It is also very interesting to notice that the EU generates more revenue for Zolgensma than does the United States. From a business perspective, this is something not seen before with such new innovative expensive therapies, especially taking into consideration that Zolgensma was launched in the EU at least a year later.

6.6.4 INNOVATIVE PAYMENT STRATEGIES TO ADDRESS PAYMENT CHALLENGES

Despite the payer and manufacturer challenges already described, there are some early successes with reimbursement of CGTs. Reasons for these early successes include relatively small patient populations (which mitigate budget impact concerns), the high level of perceived innovation and therapeutic value for patients, and innovative funding solutions that aim to address some of the uncertainty and affordability challenges expressed by payers. Some examples of the innovative approaches taken to date are outlined in Table 6.7 (229).

TABLE 6.7

Managed Entry Agreements for Gene Therapies in the European Market

Drug	United Kingdom	France	Germany	Italy	Spain
Yescarta	CDF with commercial agreements	Coverage with evidence development	Coverage with evidence development	Staged, outcome-based payment	Staged, outcome-based payment
Kymriah	CDF with commercial agreements	Coverage with evidence development	Coverage with evidence development	Staged, outcome-based payment	Staged, outcome-based payment
Zynteglo	N/A	Coverage with evidence development	Coverage with evidence development	N/A	N/A
Zolgensma	Simple discount; patient access scheme	Coverage with evidence development	Evaluation suspended	N/A	N/A

TABLE 6.7 *(Continued)*
Managed Entry Agreements for Gene Therapies in the European Market

Drug	United Kingdom	France	Germany	Italy	Spain
Tecartus	CDF with managed access scheme	Coverage with evidence development	Coverage with evidence development	N/A	N/A
Libmeldy	N/A	Coverage with evidence development	Coverage with evidence development	N/A	N/A

CDF, Cancer Drug Fund; N/A, Not available.

Outcomes-based pricing has already been implemented for a number of CGTs, with agreements reached or under consideration for Zolgensma, Luxturna, and some CAR-T therapies in certain countries. For Zolgensma, a gene therapy product for SMA, AveXis/Novartis proposed a 5-year outcome-based agreement with novel pay-over-time options. In the case of Luxturna, a gene therapy for treatment of inherited blindness, Spark Therapeutics offers outcomes-based rebates and an innovative contracting model as part of a shared risk arrangement, linking payment to both short-term efficacy and longer-term durability according to predefined outcomes. For CAR-T therapies Kymriah and Yescarta, a variety of outcomes-based models have been used across Europe, including outcomes-based rebates in Germany or outcomes-based staged payments in Italy and Spain, with pre-defined outcomes including survival and/or response rates. In France and the UK, coverage includes a requirement for evidence development, including longer-term follow-up and post-launch data from treated patients (229). Another example is Bluebird Bio, which has discussed offering an installment plan for its gene-replacement therapy Zynteglo, a therapy for β-thalassemia approved in May 2019 in the EU (240).

Models that address uncertainty as it relates to affordability have also emerged and either have or can be applied for CGTs. These include patient or expenditure thresholds, annuity models wherein the payment for a treatment is spread over time, so-called "Netflix models" wherein payers pay a fixed price per year to treat all patients, and insurance offerings such as the Cigna Embarc Benefit Protection program wherein US payers can provide coverage for Zolgensma and Luxturna by increasing insurance premiums for all plan members by roughly $1 per member per month, with no additional out-of-pocket costs for patients receiving the therapy. Recently, Aetna announced the launch in the United States of a designated Gene-based, Cellular, and Other Innovative Therapies (GCIT) network designed to enable access to new therapies while helping to manage their costs (229).

MIT's NEW Drug Development ParadIGmS (NEWDIGS) initiative, Harvard Pilgrim, and other Massachusetts payers are collaborating to develop a "performance-based annuity" approach to paying for Zolgensma. The pilot would include an initial payment and subsequent annual payments thereafter, the price of which would be determined based on how patients respond to therapy over time (231).

There are several types of payment models; the four most frequently cited models are explained here:

- *Installment-based payment.* The payer agrees to pay a fixed price for the therapy but pays in regular installments, like with an annuity, spreading the cost over a defined time.
- *Outcomes-based payment.* The payer pays only a portion of the full price up-front. If the therapy achieves pre-specified outcomes, the payer pays the remainder in full. This model spreads the risk between the payer and manufacturer.
- *Outcomes-based rebate.* The payer pays the full price of the drug up-front but receives a rebate if the drug does not achieve pre-specified outcomes. This model, again, spreads the risk between the payer and manufacturer.
- *Outcomes-based installments.* The payer pays a fixed price, with payments spread over many installments, but only if the drug continues to meet certain pre-specified outcomes. This model, too, spreads the risk between the payer and manufacturer.

Apart from managed entry agreements (MEAs), a number of innovative payment strategies have been proposed to address the high up-front budget impact of CGTs (Table 6.8). They are outlined next.

- **Amortization**

Amortization is an accounting technique that allows for the writing down of an intangible asset on the balance sheet and for the value to be split over a pre-defined

---

TABLE 6.8

Feasibility and Payer Impact of Various Payment Models Designed to Address High Up-Front Costs

Payment Model	Feasibility	Payer Impact
Amortization	Amortization can be applied to the cost of the CGT as intangible assets when generally accepted accounting principles (GAAP) are adapted and evolved. It will require changes in GAAP and International Financial Reporting Standards (IFRS) to become applicable.	Amortization allows for the high up-front budget impact of CGTs to be addressed and mitigated. Amortization allows for the CGT price to be split over several years, rather than be absorbed in the first year of acquisition and administration.

TABLE 6.8 *(Continued)*
Feasibility and Payer Impact of Various Payment Models Designed to Address High Up-Front Costs

Payment Model	Feasibility	Payer Impact
Depreciation	Depreciation is not a feasible payment model strategy, as CGTs are not considered tangible assets by payers. Once a CGT is administered, it is no longer tangible. Depreciation is already applied in the healthcare context with regard to hospitals, healthcare buildings, property, large furniture, and equipment, which are all tangible assets.	N/A
Considering CGT as an "Intangible Service"	Considering CGTs as a core element of payers' duties to cover healthcare intervention makes it difficult to consider CGTs as services.	It remains a question for payers to consider if a therapy—more specifically, a CGT—may be considered as a service that is not core to their business. It does not seem to be the case so far.
Health Leasing	Health leasing is not feasible. This is because once administered, a CGT cannot be returned, whereas in commonplace lease agreements, an asset must be returned upon termination of the lease contract. Therefore, even if this is considered a possible payment model strategy, a payer would not be able to terminate the lease once initiated.	N/A
Subscription on a Yearly Basis	As with health leasing, a subscription would not be feasible for CGTs, for the similar reason that once a subscription ends, services expected are meant to cease as well; however, once administered, CGTs cannot be returned or transferred, and neither can their effects be stopped. While leasing is payment for the use of the CGT, a subscription is an agreement for payment in advance of expected services or outcomes; in both cases neither agreements can be stopped, as the CGTs, once administered cannot be traded, transferred, returned, reversed, etc.	N/A

(Continued)

TABLE 6.8 *(Continued)*

Feasibility and Payer Impact of Various Payment Models Designed to Address High Up-Front Costs

Payment Model	Feasibility	Payer Impact
Reinsurance	If payers consider the risk of disproportional distribution of patients with highly costly therapies among insurers to be a serious issue and CGTs are eligible for reinsurance, this payment model would be feasible.	If CGTs will be eligible to be a part of reinsurance plans, payers will experience a level of protection in unpredictable distributions of the number of patients with highly costly CGTs within their plan. This would be of interest when the number of CGTs and the number of insured patients become very large. Under these circumstances, the risk of disproportional distribution of CGTs is higher.
Healthcoin or Third Party	Healthcoin, similar to reinsurance, will be most appealing between payers in a multipayer system. It could be feasible once clear definitions and conditions are outlined and abided by all payers engaging in a market utilizing healthcoins. It is unclear if current legal frameworks in several countries would allow for healthcoin to be implemented. This remains to be confirmed.	Payers, especially in a multipayer system, such as in the USA, may find this appealing, as they will continue to benefit once a patient, for whom they had provided a CGT, changes health insurance providers. This payment model strategy will not address cash flow or the high up-front budget impact. It does address the disincentive to charge a highly costly therapy where it is perceived that the value may benefit alternative payers.
Consumer Loan	Consumer loans are feasible. However, it is not clear whether certain conditions or minimal requirements must be met for a patient to be eligible to secure such a loan. Therefore, there will remain a gap of patients unable to access such a costly therapy. In the case where a patient is able to secure such a loan, payments will cease immediately upon the death of the patient. This means if a patient passes away prematurely, the remainder of the loan will remain unpaid.	This payment model does not impact the payer, payer's cash flow, or payer's budget impact, as this loan and type of payment model occur outside the traditional healthcare context. Therefore, it does not contribute to the sustainability of the healthcare system.
Payer Loan from Private or Government Organization	This payment model could be feasible. It is unclear what would happen in a multipayer system and whether payment and loan responsibility would switch to a	In the context of a payer loan, cash flow is addressed. However, when it comes to the high up-front budget impact, it will not work, as so far drugs are not amortizable.

TABLE 6.8 *(Continued)*

Feasibility and Payer Impact of Various Payment Models Designed to Address High Up-Front Costs

Payment Model	Feasibility	Payer Impact
	patient's new provider. If the GAAP and IFRS remain unchanged, drugs will not be amortizable, in which case the drug price will have to be written on the budget of the year of acquisition. This payment model will address cash flow issues.	If pharmaceuticals become amortizable, such a loan could be amortized. If this is a loan occurring in a single-payer system—the payer may already be the government with their budget already determined by the central government itself. If this type of payment model is taking place in a multipayer system, such as the USA, it is also unclear how the responsibility of such a loan would shift once a patient shifts providers.
Special Dedicated Governmental Fund	It is feasible and already implemented in single-payer systems. It may, however, prove to be more complex to implement in a fragmented payer system, such as in the USA. It may be feasible for the Centers of Medicare & Medicaid Services, if a bill was passed to allow for such a fund to be established. It may be complicated to secure the passing of such a bill, as resistance is high to increasing public intervention in healthcare.	For payers, this payment model allows for them to avoid charging the cost of expensive therapeutics while still being able to provide access to patients.
Insurance Pool	Insurance pooling is feasible and is already implemented, as shown in Germany where payers have created such funds.	Payers mutualize their funds and secure against any negative impact of disproportional distribution of patients requiring CGTs.

number of successive years according to the amortization schedule This allows one to avoid the cost of the asset being concentrated on the year of acquisition.

• **Depreciation**

Depreciation, similar to amortization, is an accounting technique applied to tangible assets and takes into account the useful life of an asset.

• **Leasing Health**

Leasing health would mean that a CGT would be paid for by payers for its use, with payments made to the CGT manufacturer on an agreed-upon periodic basis.

While leasing is common in the healthcare context, for example, with large medical equipment (such as magnetic resonance imaging machines and surgical robots), which are leased based on their use, its application to CGTs does not fit.

- **Subscription on a yearly basis**

If CGTs were subscribed to, a payer would pay fixed costs on a regular basis, for example, annually. Similar to leasing, once administered, CGTs cannot be returned and cannot be discontinued; therefore, the subscription cannot be discontinued. Furthermore, as subscriptions are based on the principle of payment in advance of expected benefits and may be used at different rates by different clients, this cannot be applied to the concept of CGTs.

- **Reinsurance**

Reinsurance occurs when payers insure themselves in the case of large, unpredictable, emergent pay-outs. Reinsurance payments could occur annually or via an agreed-upon timeline.

- **Healthcoin or Third Party**

As a new tradeable currency, healthcoin would convert incremental outcomes produced by a CGT to a common currency. This would appeal to a multipayer system where there is a high insurance provider turnover. If a payer were to pay for a CGT and the patient were to switch insurance providers, the second or following insurance provider would pay the first for the patient taken on, dependent on healthcoins.

- **Consumer Loan**

In this payment model, consumers—the patients—are responsible for securing a loan, sometimes referred to as a healthcare loan (HCL), in order to fund their costly therapy. Such a loan could also be amortized, making it more accessible for patients to receive the costly therapy.

- **Payer Loan from Private or Government Organization**

Similar to consumer loans, payers may also receive loans to fund costly therapies. Payers would be expected to pay back these loans. Payers could receive the loans via various credit mechanism sources, including the government. These loans could also be amortized, as long as the CGTs are considered amortizable.

- **Special Dedicated Governmental Fund**

In this payment model case, a government may decide to create special funds to finance CGTs. Such funds exist already, for example, the Cancer Drug Fund,

Innovative Medicines Fund, and funding based on diagnosis-related group. Such funds are usually established in single-payer systems with their budget established as either in addition to or separate from the overall health insurance budget.

- **Insurance Pool**

This payment model is characterized by several or all health insurers in a catchment area or a country teaming up to contribute to a joint fund in order to finance specific costly projects, in this context, costly CGTs.

6.6.5   Obstacles in Implementing Innovative Payment Strategies

MEAs, such as outcomes-based agreements, have been the most frequent alternative payment models used in the pharmaceutical sector so far. They may involve the collection of new evidence post-launch, e.g., from the ongoing clinical trials, to address the evidence gap in the initial assessment. For CAR-T cell therapies, including Yescarta and Kymriah, some countries (e.g., France) required that the data also be collected with existing chemotherapy data sets and national registries for bone marrow transplant or CAR-T products. Italy and Spain have a national web-based platform for which bespoke data collection requirements are created for each medicine/therapeutic indication. In Spain, the platform is being established and was not fully functional for CAR-T cell therapies (243). Moreover, there are growing concerns surrounding the fulfilment of post-market obligations. It was indicated that most post-market studies required by regulators were completed with a substantial delay and showed methodological discrepancies over time (29, 30). Information relating to randomization methods, comparator types, outcomes, and patient numbers were not sufficiently reported (31, 32). One of the most important reasons is the lack of incentives for manufacturers to collect post-launch evidence once products are on the reimbursement list. This is linked to the problem of delisting a new technology once patients and clinicians have access to and are familiar with it (244, 245).

As a result, it is unclear to know how MEAs achieved their purpose in real practice. Although MEAs provided the short-term advantage of covering new medicines at lower confidential prices, they may have limited impacts on reducing uncertainty around the comparative effectiveness and cost-effectiveness of the investigated products (246). Neyt et al. evaluated the use of MEAs in Belgium and concluded that while MEAs have a positive impact on early access to products, there are some risks in its implementation (245).

6.7   CONCLUSIONS

Manufacturers of CGTs need to establish a different business model than traditional pharmaceutical and biotechnology companies. Innovative support models to serve as a part of the care team and provide services to patients are needed and require working with regulators to reconsider the rules rather than adopting old medical-commercial models. The lifecycle of products is expected to be much shorter than conventional therapies. Therefore, the yearly return on investment needs to be significantly higher, and the development of successors has to occur more rapidly to ensure sustainability of the revenue. The lack of patentability of such therapies results in a potential rapid competition between products that are hard to differentiate. This requires an optimization of time to market and the ability to innovate constantly through better knowledge of the product value and providing services to maintain competition. Engaging cross-functional teams early in development will ensure fast, cost-efficient, and scalable processes. Industry and non-profit organizations are well-positioned to work across stakeholders to create change. Moreover, patients, caregivers, patient organizations, and healthcare professionals must work together to fight for progress. Payers and manufacturers must build systems to align prices to short- and long-term value by tracking patient outcomes over time, which will likely require cross-payer collaboration and cross manufacturers. With many CGTs targeting rare diseases and oncology, coming together to push for these technologies is even more important to ensure that manufacturers, regulatory bodies, and payers have these issues on their priority agenda.

7 Summary of Recommendations and Perspectives for Future Research

7.1 SUMMARY OF KEY FINDINGS

RMs represented substantial therapeutic promises for severe diseases lacking effective treatments. However, the complexity and novelty of RMs have raised barriers in the clinical trials, regulatory, health technology assessment (HTA) market access, and commercial activities.

- **Challenges in clinical trials:** High unmet medical needs have been important contributors to the accelerated approval of these transformative therapies based on limited evidence indicating short-term benefits outweigh possible risks. In return, regulators are expecting that post-launch evidence collection could address uncertainties regarding very long-term effectiveness and safety. However, more efforts are needed to ensure that post-market scientific obligations will be fulfilled and the post-market RWE will be comprehensive enough to validate the claimed benefits and full understanding of the value of RMs.
- **Challenges in HTA:** In general, the current HTA system may place RMs at a disadvantage considering the HTA system is not designed to fully accommodate the specialty of RMs. The substantial uncertainties in the comparative effectiveness, durability of clinical benefits, and potential future safety issues constituted the biggest challenge for the HTA of RM. Furthermore, limitations in clinical evidence undermine the robustness of the economic analysis, which must rely on intensive extrapolations and assumptions on the long-term outcomes. This raises a high uncertainty for the economic evaluation outcomes for RMs.
- **Challenges in affordability:** The high prices of RMs raised controversies regarding whether such prices are justified in relation to the manufacturing cost and clinical benefits. Payers will face a financial crisis to cover upcoming RMs targeting not only rare diseases but also more prevalent diseases. The constricted payers' budget and divergence in the coverage policies will possibly lead to unequal access to RMs. Outcome-based or finance-based agreements are increasingly used for RMs,

DOI: 10.1201/9781003366676-7

while there are practical and legislative hurdles to overcome to allow the broader use of these agreements. However, such outcomes will not address the key challenge of the budget impact. So far only amortization of RMs may allow the control of the budget impact on such therapies.

- **Disruptions caused by COVID-19:** All these challenges have been and may in the future be amplified by the COVID-19 pandemic, which urged manufacturers of cell and gene therapies (CGTs) to reshape their clinical development plans, business models, and market access strategies.

The importance of partnerships: The successful market access of advanced therapy medicinal products (ATMPs) could not be accomplished by one sole party; instead, it requires intensive engagements of all relevant stakeholders across different agencies and even across countries. Moreover, the collaborations of different players with complementary capacities will be the key to expedite the transition from scientific discovery to the ultimate commercialization. It remains to be seen if in the future large pharmaceutical companies will continue to partner or will start to integrate in-house the whole value chain of RMs.

7.2   RECOMMENDATIONS FOR MANUFACTURERS

7.2.1   PARTNERSHIPS TO STREAMLINE THE MARKET ACCESS OF RMS

- **External collaborations**

The materialization of the commercial potential of RMs will rely on an "innovative ecosystem," a collaborative arrangement through which each individual company combines its individual experiences into a coherent solution. The core element of an innovative ecosystem is that it leverages synergies across multiple companies/organizations to bring value to customers that no company/organization alone could deliver (212). Academia, not-for-profit organizations (e.g., patient group or charities), biotechnology companies, and service companies will continue to intricately collaborate in both upstream and downstream activities, spanning from early research to achieving developmental milestones, which will promote the value of assets and attract well-financed partners. Examples of collaborations could include:

- Partnerships with multiple suppliers of raw materials: Manufacturing remains a rate-limiting factor in the production of CGTs. Finding ways to address bottlenecks in their current manufacturing processes is crucial to ensure business continuity and improve resilience in coping with future disruptions. While more viral vector manufacturing capacity is the long-term answer, CGT companies might consider strategic partnerships with key suppliers of critical raw materials and viral vectors, or identify and validate two to three potential suppliers early in the development process, rather than relying on just one (154).

- "Point of care" delivery approach: This could be a more resilient manufacturing and delivery model that possibly brings the manufacturing and delivery of therapies to a single site—the point of care (247).
- Prepare for digitalization: Digitalization is expected to reduce the time needed to translate the CGTs from laboratory tests to commercial use (166). For example, digitalization could be utilized to promote virtual audits if risk-based approaches are taken, and the audit partner would have the ability to send documentation in an efficient way (166). Companies that are beginning to scale up manufacturing as their businesses grow may also reconsider the criteria for selecting manufacturing partners. Reliability and preparedness for digitalization may be valued more highly in the current crisis when regulatory site inspections are restricted (154).

7.2.2 Strategies to Strengthen the Quality of Clinical Evidence

- **Alternative control group for single-arm trials**
  - Indirect comparisons with evidence generated from natural history studies, external observational studies, or patient registries
  - Suitable methods to mitigate the potential confounding factors and patient heterogeneity in the external comparison groups
- **Innovative study design to cope with small patient size**
  - Adaptive study design to allow the pre-defined modifications along with the evidence accumulation
  - Other strategies: Bayesian analysis to enhance the precision of study results by aggregating all the evidence available, simulation/modeling to extrapolate the short-term evidence, and composite endpoints to increase the rate of event and statistic efficiency
- **Validation of surrogate endpoints**
  - Define and report surrogate endpoints correctly to enhance better interpretation of study results and enable between-studies comparison
  - Conduct validity studies to establish the relationship between surrogate endpoints and patient-relevant endpoints
  - Identify reliable biomarkers relevant to clinical final outcomes through natural history studies
- **Telemedicine and clinical trials**
  - Telemedicine will be powerful to encourage patient recruitment and improve adherence to follow-up schedules (166). This will be particularly relevant for clinical trials of CGTs, through which the pool of patients will increase because patients, usually very sick or disabled, will not need to travel long distances to specialized medical centers for participation and follow-up in the trial (186).
  - Although feasible, it should be noted that remote approaches may increase within-subject variability in the collection of patient-reported outcomes (PROs); thus the preferred collection method (e.g., email,

telephone, and internet) will depend on the specifics of the clinical questions to be answered and the PRO instruments used (159).

7.2.3 ACTIVE ENGAGEMENTS WITH OTHER STAKEHOLDERS

• **Patient involvement**

Patient involvement plays a crucial role throughout the whole lifecycle of developing ATMPs or rare diseases. In the stage of clinical trials, early interactions with patient advocacy groups are beneficial to obtain in-depth knowledge about the characteristics of rare diseases, to reach the target patient population quickly, to promote the proper design and conduction of clinical trials, and to facilitate the collection of patient opinions. In the HTA process, patients could provide valuable insights regarding the PROs measuring patients' daily living activities, and QOL can provide powerful evidence to understand patient perceptions on what matters most to them. The impact of patient engagement to bring financial value by facilitating enrollment, improving adherence, and avoiding protocol amendments should be amplified (248).

• **Healthcare providers**

Manufacturers should work closely with healthcare providers to ensure that appropriate infrastructures and operation standards for ATMP delivery have been established. This is one critical factor impacting claimed clinical benefits of ATMPs that could be observed in the real-world setting. In the HTA process, given that many ATMPs targeting inherited rare diseases have the onset of symptoms in childhood, clinician-reported outcomes (CliniRO) would be prioritized for children too young to reliably self-report their symptoms and experiences (249). Furthermore, consultations with clinical experts on the design and build of economic models are important in order to obtain their insights into appropriate input parameters where published data are lacking and to ensure that key assumptions are representative of clinical practice (250).

• **Other RM developers for rare diseases**

A global network aggregating scarce resources is essential to enhance the sharing of research data under the principles of open science with publicly accessible resources, tools, and knowledge bases (251). Global data sharing for natural history studies across research networks or pharmaceutical companies could be efficient to minimize duplicative works and reduce the research burdens on the small patient community (252).

• **Parallel dialogue with regulators and payers**

There is an urgent call for enhancing interactions and communications between regulators and payers in order to ensure that 1) all the evidence uncertainties (e.g., surrogate outcomes) with the innovative ATMPs have been sufficiently communicated, 2) agreements are reached regarding the drugs' qualifications for expedited approval programs based on the "unmet clinical needs" principle, and 3) the evidence requirements for pivotal trials supporting market approval, as well as qualities of confirmatory studies in the post-marketing obligations will meet the expectations of both parties (Figure 7.1).

7.3 IMPLICATIONS FOR FUTURE RESEARCH EFFORTS

7.3.1 FROM THE PERSPECTIVE OF MARKET AUTHORIZATION

- **More clarifications and more transparency in the regulatory activities**
  - To standardize the definitions of "unmet needs" to ensure that only qualified products will benefit from expedited programs
  - To clarify the requirements for using surrogate endpoints as primary endpoints
  - To increase the transparency of the consultation activities provided to manufacturers
- **Ensure that post-market obligations are fulfilled**
  - To strengthen the mechanism of monitoring to guarantee that post-market studies proceed as planned
  - To explore the potential of building up centralized patient registries to facilitate the post-market evidence collection of efficacy and safety data
  - To highlight the importance of coordination across decision-makers (i.e., regulators and payers) and collaboration across countries to improve the consistency and efficiency of post-launch RWE collection
- **International harmonization for RM regulations**
  - To standardize the terminology and to establish a universal regulatory pathway for RMs to align evidence requirements on the highest standard and allow international development to accelerate global access to innovative therapy

7.3.2 FROM THE PERSPECTIVE OF HTA AND AFFORDABILITY

- **Methodology considerations**
  - To make assessments of the ethical and social aspects of RMs in a robust way by resorting to experts from multiple disciplines
  - To examine the rationality of adjusting the principle of current reference cases, such as implementing a flexible ICER threshold and different discount rules for RMs
  - To standardize the methods of defining and collecting RWE
  - To enhance the capacities of evaluating and incorporating RWE in the HTA decision-making process

Pre-clinical

- Collaborate with academia for the proof-of-concept validation
- Collaborate with other biopharmaceutical companies with requisite manufacturing materials or infrastructures
- Research networks for sharing knowledge and resources with other developer
- Engage with patients and physicians to understand the unmet clinical needs to be addressed
- Communicate with regulators to discuss the potential for accelerated programs (e.g., orphan drug)

Clinical trials

- Engage with patient groups to understand the disease natural history, to support the patient recruitment, and to improve the compliance with the clinical trial protocol
- Early dialogues with regulators and HTA bodies to obtain insights on the evidence requirements to be fulfilled for clinical trials
- Seek scientific advices and parallel consultation on the design of pivotal studies and understand their acceptance of innovative methodology

Market authorization

- Interaction with regulators to discuss the eligibility for expedited approval pathways
- 'Biotech' collaborates with 'Big pharma' to strengthen their capacities for navigating the regulatory activities, or transfer the regulatory-related responsibility to 'Big pharma'
- Global regulatory collaboration to standardize the terminology and clarify the minimal evidence for market approval of ATMPs

HTA and reimbursement

- Patient involvement to obtain their perceptions on the potential values of ATMPs, such as via collection of QoL evidence and other patient reported outcomes
- Physician involvement to offer expert inputs when the reliable inputs for economic analysis are absent from previous publications
- Communications with payers in induvial country for the preparation of tailored reimbursement dossier

Post-market

- Communications with the regulators and HTA bodies to ensure that study design of post-market studies would satisfy the requirements of both parties
- Engage with patient organizations to increase patient enrolment in the patient registries and enable close patient follow-up and monitoring
- Provide physicians training to make sure that sufficient infrastructures for administration and management of ATMPs are in place, as well as enhancing their knowledge on ATMPs to increase clinical adoption

FIGURE 7.1 Collaborations in each phase of development.

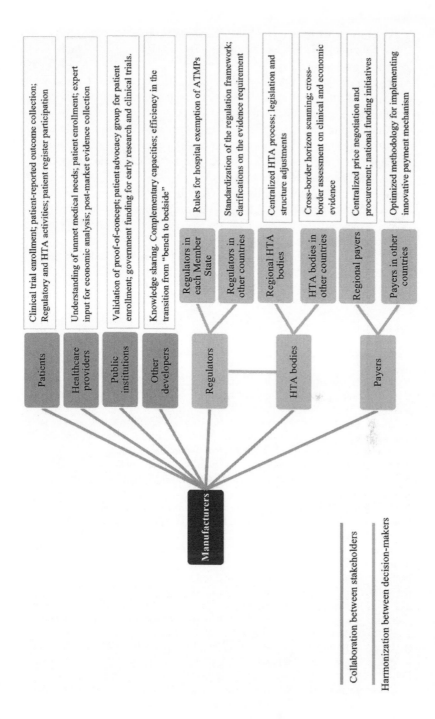

FIGURE 7.2 Collaborations between all stakeholders to facilitate market access.

- **Alternative payment strategies**
  - To ensure that relevant and realistic clinical targets are set for outcome-based payment, and payment terms are mutually beneficial to manufacturers and payers.
  - To examine whether the modifications of legislation will be necessary to mitigate the challenges when patients switch plans and to avoid the adverse selection of patients.
  - Amortization remains the most appropriate way forward to absorb the initial high budget impact. However, it remains challenging to implement and will require several steps to change the generally accepted accounting principle for RMs.
- **Joint HTA and payment activities across borders**
  - To explore whether joint HTA activities across borders will close or aggravate the existing inequity between developed countries and non-developed countries (Figure 7.2)
  - To explore the appropriate governance approaches and clarified working structure for the implementation of joint activities involving multiple countries

7.4 REFERENCES

1. Chris Mason PD. A brief definition of regenerative medicine. Regenerative Medicine. 2008; 3: 1–5.
2. Buzhor E, Leshansky L, Blumenthal J, Barash H, Warshawsky D, Yaron M, et al. Cell-based therapy approaches the hope for incurable diseases. Regenerative Medicine. 2014; 9(5): 649–72.
3. Mckinsey Company. Biopharma portfolio strategy in the era of cell and gene therapy. 2020. https://www.mckinsey.com/industries/life-sciences/our-insights/biopharma-portfolio-strategy-in-the-era-of-cell-and-gene-therapy
4. Alliance for Regenerative Medicine. Advancing gene, cell, & tissue based therapies-ARM Annual Report & Sector Year in Review: 2019. 2019. https://alliancerm.org/sector-report/2019-annual-report
5. Qiu T, Hanna E, Dabbous M, et al. Regenerative medicine regulatory policies: A systematic review and international comparison. Health Policy. 2020; 124: 701–13.
6. Davies BM, Smith J, Rikabi S, et al. A quantitative, multi-national and multi-stakeholder assessment of barriers to the adoption of cell therapies. Journal of Tissue Engineering. 2017; 8: 2041731417724413.
7. Hanna E, Toumi M, Dussart C, et al. Funding breakthrough therapies: A systematic review and recommendation. Health Policy. 2018; 122: 217–29.
8. Ten Ham RMT, Hoekman J, Hovels AM, et al. Challenges in advanced therapy medicinal product development: A survey among companies in Europe. Molecular Therapy – Methods & Clinical Development. 2018; 11: 121–30.
9. Bravery CA, Ball O, Robinson S. EU market authorisation strategy: Lessons from the first 22 ATMP submitted to the EMA. Cell and Gene Therapy Insights. 2019; 5: 759–91.
10. de Wilde S, Coppens DGM, Hoekman J, et al. EU decision-making for marketing authorization of advanced therapy medicinal products: A case study. Drug Discovery Today. 2018; 23: 1328–33.

11. Berger I, Ahmad A, Bansal A, et al. Global distribution of businesses marketing stem cell-based interventions. Cell Stem Cell. 2016; 19: 158–62.

12. Davis C, Lexchin J, Jefferson T, et al. "Adaptive pathways" to drug authorisation: Adapting to industry? British Medical Journal. 2016; 354: i4437.

13. Banzi R, Gerardi C, Bertele V, et al. Approvals of drugs with uncertain benefit-risk profiles in Europe. European Journal of Internal Medicine. 2015; 26: 572–84.

14. Andrews PW, Cavagnaro J, Deans R, et al. Harmonizing standards for producing clinical-grade therapies from pluripotent stem cells. Nature Biotechnology. 2014; 32: 724–6.

15. Hofer MP, Jakobsson C, Zafiropoulos N, et al. Regulatory watch: Impact of scientific advice from the European Medicines Agency. Nature Reviews Drug Discovery. 2015; 14: 302–3.

16. Kellathur SN, Lou HX. Cell and tissue therapy regulation: worldwide status and harmonization. Biologicals. 2012; 40: 222–4.

17. Rosemann A, Vasen F, Bortz G. Global diversification in medicine regulation: Insights from regenerative stem cell medicine. Science as Culture. 2018: 1–27.

18. Rosemann A, Bortz G, Vasen F, Sleeboom-Faulkner M. Global regulatory developments for clinical stem cell research diversification and challenges to collaborations. Regenerative Medicine. 2016; 11: 647–57.

19. Sipp D. Conditional approval: Japan lowers the bar for regenerative medicine products. Cell Stem Cell. 2015; 16: 353–6.

20. Kondo H, Hata T, Ito K, et al. The current status of Sakigake designation in Japan, PRIME in the European Union, and breakthrough therapy designation in the United States. Therapeutic Innovation & Regulatory Science. 2016; 51: 51–4.

21. Nagpal A, Juttner C, Hamilton-Bruce MA, et al. Stem cell therapy clinical research: A regulatory conundrum for academia. Advanced Drug Delivery Reviews. 2017; 122: 105–14.

22. Coppens DGM, De Bruin ML, Leufkens HGM, et al. Global regulatory differences for gene- and cell-based therapies: Consequences and implications for patient access and therapeutic innovation. Clinical Pharmacology & Therapeutics. 2018; 103: 120–7.

23. Beaver JA, Howie LJ, Pelosof L, et al. A 25-year experience of US Food and Drug Administration accelerated approval of malignant hematology and oncology drugs and biologics: A review. JAMA Oncology. 2018; 4: 849–56.

24. Chen EY, Raghunathan V, Prasad V. An overview of cancer drugs approved by the US Food and Drug Administration based on the surrogate end point of response rate. JAMA Internal Medicine. 2019; 179: 915–21.

25. Hwang TJ, Franklin JM, Chen CT, Lauffenburger JC, Gyawali B, Kesselheim AS, Darrow JJ. Efficacy, safety, and regulatory approval of Food and Drug Administration–designated breakthrough and nonbreakthrough cancer medicines. Journal of Clinical Oncology. 2018; 36: 1805–12.

26. Naci H, Smalley KR, Kesselheim AS. Characteristics of preapproval and postapproval studies for drugs granted accelerated approval by the US Food and Drug Administration. JAMA. 2017; 318: 626–36.

27. Fritsche E, Elsallab M, Schaden M, et al. Post-marketing safety and efficacy surveillance of cell and gene therapies in the EU: A critical review. Cell and Gene Therapy Insights. 2019; 5: 1505–21.

28. Banzi R, Gerardi C, Bertele V, et al. Conditional approval of medicines by the EMA. British Medical Journal. 2017; 357: j2062.

29. Gyawali B, Hey SP, Kesselheim AS. Assessment of the Clinical benefit of cancer drugs receiving accelerated approval. JAMA Internal Medicine. 2019; 179: 906–13.

30. Hoekman J, Klamer TT, Mantel-Teeuwisse AK, et al. Characteristics and follow-up of postmarketing studies of conditionally authorized medicines in the EU. British Journal of Clinical Pharmacology. 2016; 82: 213–26.

31. Wallach JD, Egilman AC, Dhruva SS, et al. Postmarket studies required by the US Food and Drug Administration for new drugs and biologics approved between 2009 and 2012: cross sectional analysis. British Medical Journal. 2018; 361: k2031.

32. Makady A, van Veelen A, de Boer A, et al. Implementing managed entry agreements in practice: The Dutch reality check. Health Policy. 2019; 123: 267–74.

33. Pauwels K, Huys I, Vogler S, et al. Managed entry agreements for oncology drugs: Lessons from the European experience to inform the future. Frontiers in Pharmacology. 2017; 8: 171.

34. Hans-Georg Eichler FP, Bruno Flamion, Hubert Leufkens and Alasdair Breckenridge. Balancing early market access to new drugs with the need for benefit/risk data: A mounting dilemma. Nature. 2008; 7: 818–26.

35. European Medicines Agency. Guideline on safety and efficacy follow-up and risk management of advanced therapy medicinal products. 2018. www.ema.europa.eu/en/guideline-safety-efficacy-follow-risk-management-advanced-therapy-medicinal-products.

36. U.S. Food and Drug Administration. Long term follow-up after administration of human gene therapy products. 2018. https://www.fda.gov/regulatory-information/search-fda-guidance-documents/long-term-follow-after-administration-human-gene-therapy-products.

37. U.S. Food and Drug Administration. Statement from FDA Commissioner Scott Gottlieb, M.D. and Peter Marks, M.D., Ph.D., Director of the Center for Biologics Evaluation and Research on new policies to advance development of safe and effective cell and gene therapies. FDA Statement, 2019. https://www.fda.gov/news-events/press-announcements/statement-fda-commissioner-scott-gottlieb-md-and-peter-marks-md-phd-director-center-biologics.

38. Hanna E, Remuzat C, Auquier P, et al. Advanced therapy medicinal products: Current and future perspectives. Journal of Market Access & Health Policy. 2016; 4.

39. MTPConnect. Regenerative medicine: Opportunities for Australia. 2018. https://www.mtpconnect.org.au/Category?Action=View&Category_id=195.

40. KPMG. Following through realizing the promise of stem cells: A Canadian Stem Cell Strategy and Action Plan (2015–2025). 2014. https://ec.europa.eu/futurium/en/system/files/ged/following-through-realizing-the-promise-of-stem-cells.pdf

41. Ministry of Health. Therapeutic products regulatory scheme consultation. 2018. www.health.govt.nz/publication/therapeutic-products-regulatory-scheme-consultation.

42. International Pharmaceutical Regulators Programme (IPRP). Gene Therapy Working Group. https://www.iprp.global/working-group/gene-therapy.

43. International Council for Harmonisation of Technical Requirements for Pharmaceuticals for Human Use (ICH). Gene Therapy Discussion Group. https://www.ich.org/page/consideration-documents.

44. International Alliance for Biological Standardization (IABS). Cell and gene therapy. https://www.iabs.org/cell-gene-therapy/

45. World Health Organization (WHO). WHO Considerations on Regulatory Convergence of Cell and Gene Therapy Products. 2021. https://cdn.who.int/media/docs/default-source/biologicals/ecbs/who-public-consultation_cgtp-white-paper_16_dec_2021.pdf?sfvrsn=18f6c549_5.

46. Health Action International (HAI). Marketing authorisation flexibilities that enable early access to medicines should only respond to true unmet medical needs and must protect patients' safety. 2015. https://haiweb.org/publication/early-access-to-

medicines-marketing-authorisation-flexibilities-should-only-respond-to-unmet-medical-needs-must-protect-patient-safety/.

47. Ombudsman E. Decision in strategic inquiry OI/7/2017/KR on how the European Medicines Agency engages with medicine developers in the period leading up to applications for authorisations to market new medicines in the EU. 2019. https://www.ombudsman.europa.eu/en/decision/en/116683.

48. Michalopoulos S. Stakeholders call for more transparent meetings between pharma and EMA. 2018. https://www.euractiv.com/section/health-consumers/news/stakeholders-call-for-more-transparent-meetings-between-pharma-and-ema/.

49. Hampson G, Towse A, Dreitlein WB, Henshall C, Pearson SD. Real-world evidence for coverage decisions: Opportunities and challenges. Journal of Comparative Effectiveness Research. 2018; 7: 1133–43.

50. Drug Commission of the German Medical Association. Opinion on the "cost-sharing initiatives" and "risk-sharing agreements" between pharmaceutical manufacturers and health and hospital. 2020. https://www.deutschland.de/en/topic/life/society-integration/medicines.

51. Jansen-van der Weide MC, Gaasterland CMW, Roes KCB, et al. Rare disease registries: potential applications towards impact on development of new drug treatments. Orphanet Journal of Rare Diseases. 2018; 13: 154.

52. Chow SC, Chang YW. Statistical considerations for rare diseases drug development. Journal of Biopharmaceutical Statistics. 2019; 29: 874–86.

53. Jørgensen J, Hanna E, Kefalas P. Outcomes-based reimbursement for gene therapies in practice: the experience of recently launched CAR-T cell therapies in major European countries. Journal of Market Access & Health Policy. 2020; 15: 1715536.

54. Jørgensen J, Kefalas P. Upgrading the SACT dataset and EBMT registry to enable outcomes-based reimbursement in oncology in England: A gap analysis and top-level cost estimate. Journal of Market Access & Health Policy. 2019; 7: 1635842–42.

55. European Medicines Agency (EMA). Patient register. 2020. https://www.ema.europa.eu/en/human-regulatory/post-authorisation/patient-registries.

56. European Medicines Agency (EMA). Qualification opinion on cellular therapy module of the European Society for Blood & Marrow Transplantation (EBMT) Registry. 2019. https://www.ema.europa.eu/en/documents/scientific-guideline/qualification-opinion-cellular-therapy-module-european-society-blood-marrow-transplantation-ebmt_en.pdf.

57. Pechmann A, Konig K, Bernert G, et al. SMArtCARE — A platform to collect real-life outcome data of patients with spinal muscular atrophy. Orphanet Journal of Rare Diseases. 2019; 14: 18.

58. Ermisch M, Bucsics A, Vella Bonanno P, et al. Payers' views of the changes arising through the possible adoption of adaptive pathways. Frontiers in Pharmacology. 2016; 7: 305.

59. Angelis A, Naci H, Hackshaw A. Recalibrating health technology assessment methods for cell and gene therapies. Pharmacoeconomics. 2020; 38: 1297–308.

60. Spoors J, Miners A, Cairns J, et al. Payer and implementation challenges with advanced therapy medicinal products (ATMPs). BioDrugs. 2020; 35: 1–5.

61. Hampson G, Towse A, Pearson SD, et al. Gene therapy: Evidence, value and affordability in the US health care system. Journal of Comparative Effectiveness Research. 2018; 7: 15–28.

62. Lloyd-Williams H, Hughes DA. A systematic review of economic evaluations of advanced therapy medicinal products. British Journal of Clinical Pharmacology. 2020.

63. Viriato D, Bennett N, Sidhu R, et al. An economic evaluation of voretigene neparv-ovec for the treatment of biallelic RPE65-mediated inherited retinal dystrophies in the UK. Advances in Therapy. 2020; 37: 1233–47.

64. de Windt TS, Sorel JC, Vonk LA, et al. Early health economic modelling of single-stage cartilage repair. Guiding implementation of technologies in regenerative medicine. Journal of Tissue Engineering and Regenerative Medicine. 2017; 11: 2950–9.

65. Roth JA, Sullivan SD, Lin VW, et al. Cost-effectiveness of axicabtagene ciloleucel for adult patients with relapsed or refractory large B-cell lymphoma in the United States. Journal of Medical Economics. 2018; 21: 1238–45.

66. South E, Cox E, Meader N, et al. Strimvelis((R)) for treating severe combined immunodeficiency caused by adenosine deaminase deficiency: An evidence review group perspective of a NICE highly specialised technology evaluation. PharmacoEconomics Open. 2019; 3: 151–61.

67. Drummond MF, Neumann PJ, Sullivan SD, et al. Analytic considerations in apply-ing a general economic evaluation reference case to gene therapy. Value in Health. 2019; 22: 661–8.

68. Ribera Santasusana JM, de Andres Saldana A, Garcia-Munoz N, et al. Cost-effectiveness analysis of tisagenlecleucel in the treatment of relapsed or refrac-tory B-cell acute lymphoblastic leukaemia in children and young adults in Spain. ClinicoEconomics and Outcomes Research. 2020; 12: 253–64.

69. Senior M. Rollout of high-priced cell and gene therapies forces payer rethink. Nature Biotechnology. 2018; 36: 291–2.

70. Driscoll D, Farnia S, Kefalas P, et al. Concise review: The high cost of high tech medicine: Planning ahead for market access. Stem Cells Translational Medicine. 2017; 6: 1723–9.

71. Aballéa S, Thokagevistk K, Velikanova R, et al. Health economic evaluation of gene replacement therapies: methodological issues and recommendations. Journal of Market Access & Health Policy. 2020; 8: 1822666.

72. Walton M, Sharif S, Simmonds M, et al. Tisagenlecleucel for the treatment of relapsed or refractory B-cell acute lymphoblastic leukaemia in people aged up to 25 years: An evidence review group perspective of a NICE single technology appraisal. Pharmacoeconomics. 2019; 37: 1209–17.

73. Goncalves E. Advanced therapy medicinal products: Value judgement and ethical evaluation in health technology assessment. European Journal of Health Economics. 2020; 21: 311–20.

74. Qiu T, Hanna E, Dabbous M, et al. Health technology assessment of gene therapies in Europe and the USA: Analysis and future considerations. Cell and Gene Therapy Insights. 2019; 5: 1043–59.

75. Furzer J, Gupta S, Nathan PC, et al. Cost-effectiveness of tisagenlecleucel vs stan-dard care in high-risk relapsed pediatric acute lymphoblastic leukemia in Canada. JAMA Oncology. 2020.

76. Jonsson B, Hampson G, Michaels J, et al. Advanced therapy medicinal products and health technology assessment principles and practices for value-based and sustain-able healthcare. European Journal of Health Economics. 2019; 20: 427–38.

77. Lin JK, Muffly LS, Spinner MA, et al. Cost effectiveness of chimeric antigen recep-tor T-cell therapy in multiply relapsed or refractory adult large B-cell lymphoma. Journal of Clinical Oncology: Official Journal of the American Society of Clinical Oncology. 2019; 37: 2105–19.

78. Petrou P. Is it a Chimera? A systematic review of the economic evaluations of CAR-T cell therapy. Expert Review of Pharmacoeconomics & Outcomes Research. 2019; 19: 529–36.

79. Prasad V. Immunotherapy: Tisagenlecleucel — the first approved CAR-T-cell therapy: implications for payers and policy makers. Nature Reviews Clinical Oncology. 2018; 15: 11–2.

80. Retel VP, Steuten LMG, Geukes Foppen MH, et al. Early cost-effectiveness of tumor infiltrating lymphocytes (TIL) for second line treatment in advanced melanoma: A model-based economic evaluation. BMC Cancer. 2018; 18: 895.

81. Gavan SP, Lu CY, Payne K. Assessing the joint value of genomic-based diagnostic tests and gene therapies. Journal of Personalized Medicine. 2019; 9.

82. Johnson S, Buessing M, O'Connell T, et al. Cost-effectiveness of voretigene neparvovec-rzyl vs standard care for RPE65-mediated inherited retinal disease. JAMA Ophthalmology. 2019; 137: 1115–23.

83. Cook K, Forbes SP, Adamski K, et al. Assessing the potential cost-effectiveness of a gene therapy for the treatment of hemophilia A. Journal of Medical Economics. 2020; 23: 501–12.

84. Whittington MD, McQueen RB, Campbell JD. Valuing chimeric antigen receptor T-cell therapy: Current evidence, uncertainties, and payment implications. Journal of Clinical Oncology: Official Journal of the American Society of Clinical Oncology. 2020; 38: 359–66.

85. Nagpal A, Milte R, Kim SW, et al. Economic evaluation of stem cell therapies in neurological diseases: A systematic review. Value in Health. 2019; 22: 254–62.

86. Lin JK, Lerman BJ, Barnes JI, et al. Cost effectiveness of chimeric antigen receptor T-cell therapy in relapsed or refractory pediatric B-cell acute lymphoblastic leukemia. Journal of Clinical Oncology: Official Journal of the American Society of Clinical Oncology. 2018; 36: 3192–202.

87. Zimmermann M, Lubinga SJ, Banken R, et al. Cost Utility of voretigene neparvovec for biallelic RPE65-mediated inherited retinal disease. Value in Health. 2019; 22: 161–7.

88. Thielen FW, van Dongen-Leunis A, Arons AMM, et al. Cost-effectiveness of Anti-CD19 chimeric antigen receptor T-Cell therapy in pediatric relapsed/refractory B-cell acute lymphoblastic leukemia. A societal view. European Journal of Haematology. 2020; 105: 203–15.

89. Hettle R, Corbett M, Hinde S, et al. The assessment and appraisal of regenerative medicines and cell therapy products: an exploration of methods for review, economic evaluation and appraisal. Health Technology Assessment. 2017; 21: 1–204.

90. Raymakers AJN, Regier DA, Peacock SJ. Modelling uncertainty in survival and cost-effectiveness is vital in the era of gene therapies: The case of axicabtagene ciloleucel. Health Policy and Technology. 2019; 8: 103–4.

91. Fiorenza S, Ritchie DS, Ramsey SD, et al. Value and affordability of CAR T-cell therapy in the United States. Bone Marrow Transplantation. 2020; 55: 1706–15.

92. Skinner MW. Gene therapy for hemophilia: Addressing the coming challenges of affordability and accessibility. Molecular Therapy. 2013; 21: 1–2.

93. Buessing M, O'Connell T, Johnson S, et al. Important considerations in modeling the cost-effectiveness for the first food and drug administration-approved gene therapy and implications for future one-time therapies. Value in Health. 2019; 22: 970–1.

94. Ginty PJ, Singh PB, Smith D, et al. Achieving reimbursement for regenerative medicine products in the USA. Regenerative Medicine. 2010; 5: 463–9.

95. McGrath E, Chabannon C, Terwel S, et al. Opportunities and challenges associated with the evaluation of chimeric antigen receptor T cells in real-life. Current Opinion Oncology. 2020; 32: 427–33.

96. Jorgensen J, Hanna E, Kefalas P. Outcomes-based reimbursement for gene therapies in practice: The experience of recently launched CAR-T cell therapies in major European countries. Journal of Mark Access & Health Policy. 2020; 8: 1715536.

97. Carr DR, Bradshaw SE. Gene therapies: the challenge of super-high-cost treatments and how to pay for them. Regenerative Medicine. 2016; 11: 381–93.

98. Schaffer SK, Messner D, Mestre-Ferrandiz J, et al. Paying for cures: Perspectives on solutions to the "affordability issue". Value in Health. 2018; 21: 276–9.

99. Flowers CR, Ramsey SD. What can cost-effectiveness analysis tell us about chimeric antigen receptor T-cell therapy for relapsed acute lymphoblastic leukemia? Journal of Clinical Oncology: Official Journal of the American Society of Clinical Oncology. 2018: Jco2018793570.

100. Machin N, Ragni MV, Smith KJ. Gene therapy in hemophilia A: A cost-effectiveness analysis. Blood Advances. 2018; 2: 1792–8.

101. Whittington MD, McQueen RB, Ollendorf DA, et al. Long-term survival and cost-effectiveness associated with axicabtagene ciloleucel vs chemotherapy for treatment of B-cell lymphoma. JAMA Network Open. 2019; 2: e190035.

102. Zhu F, Wei G, Zhang M, et al. Factors associated with costs in chimeric antigen receptor T-cell therapy for patients with relapsed/refractory B-cell malignancies. Cell Transplantation. 2020; 29: 0963689720919434.

103. Champion AR, Lewis S, Davies S, et al. Managing access to advanced therapy medicinal products: Challenges for NHS Wales. British Journal of Clinical Pharmacology. 2020: 1–6.

104. Roth JA, Sullivan SD, Lin VW, et al. Cost-effectiveness of axicabtagene ciloleucel for adult patients with relapsed or refractory large B-cell lymphoma in the United States. Journal of Medical Economics. 2018; 21: 1238–45.

105. Jorgensen J, Servos S, Kefalas P. The potential price and access implications of the cost-utility and budget impact methodologies applied by NICE in England and ICER in the US for a novel gene therapy in Parkinson's disease. Journal of Mark Access & Health Policy. 2018; 6: 1500419.

106. Institute for Clinical and Economic Review (ICER). Adapted value assessment methods for high-impact "Single and Short-Term Therapies" (SSTs). 2019. https://icer-review.org/announcements/final_potential_cures_methods/.

107. Waldeck AR, Botteman MF, White RE, et al. The importance of economic perspective and quantitative approaches in oncology value frameworks of drug selection and shared decision making. Journal of Managed Care & Specialty Pharmacy. 2017; 23: S6–12.

108. Pettitt DA, Raza SD, Smith J. The limitations of QALY: A literature review. Journal of Stem Cell Research & Therapy. 2016; 06.

109. Wolowacz SE, Briggs A, Belozeroff V, et al. Estimating health-state utility for economic models in clinical studies: An ISPOR good research practices task force report. Value in Health. 2016; 19: 704–19.

110. White W. A rare disease patient/caregiver perspective on fair pricing and access to gene-based therapies. Gene Therapy. 2019; 27: 474–81.

111. Sara Silbert GAY, Andrew G. Shuman. How should we determine the value of CAR T-cell therapy? AMA Journal of Ethics. 2019; 21: E844–51.

112. Cho E, Yoo S-L, Kang Y, et al. Reimbursement and pricing of regenerative medicine in South Korea: key factors for achieving reimbursement. Regenerative Medicine. 2020; 15: 1550–60.

113. Mahalatchimy A. Reimbursement of cell-based regenerative therapy in the UK and France. Medical Law Review. 2016; 24: 234–58.

114. Garrison LP, Jackson T, Paul D, et al. Value-based pricing for emerging gene therapies: The economic case for a higher cost-effectiveness threshold. Journal of Managed Care & Specialty Pharmacy.2019; 25: 793–9.

115. Yeung K, Suh K, Garrison LP, Jr., et al. Defining and managing high-priced cures: Healthcare payers' opinions. Value in Health. 2019; 22: 648–55.

116. Husereau D. How do we value a cure? Expert Review of Pharmacoeconomics & Outcomes Research. 2015; 15: 551–5.

117. Faulkner E, Spinner DS, Ringo M, et al. Are global health systems ready for transformative therapies? Value in Health. 2019; 22: 627–41.

118. Towse A, Fenwick E. Uncertainty and cures: Discontinuation, irreversibility, and outcomes-based payments: what is different about a one-off treatment? Value in Health. 2019; 22: 677–83.

119. Salzman R, Cook F, Hunt T, et al. Addressing the value of gene therapy and enhancing patient access to transformative treatments. Molecular Therapy. 2018; 26: 2717–26.

120. Touchot N, Flume M. The payers' perspective on gene therapies. Nature Biotechnology. 2015; 33: 902–4.

121. de Lima Lopes G, Nahas GR. Chimeric antigen receptor T cells, a savior with a high price. Chinese Clinical Oncology. 2018; 7: 21.

122. Cho E YS, Kang Y, Lee JH. Reimbursement and pricing of regenerative medicine in South Korea: key factors for achieving reimbursement. Regenerative Medicine. 2020; Epub ahead of print.

123. Champion AR, Lewis S, Davies S, et al. Managing access to advanced therapy medicinal products: Challenges for NHS Wales. British Journal of Clinical Pharmacology. 2020; 87(6): 2444–49.

124. Barlow JF, Yang M, Teagarden JR. Are payers ready, willing, and able to provide access to new durable gene therapies? Value in Health. 2019; 22: 642–7.

125. Patel N, Farid SS, Morris S. How should we evaluate the cost-effectiveness of CAR T-cell therapies? Health Policy and Technology. 2020; 9(3): 271–73.

126. Rose JB, Williams DJ. The UK relative to other single payer-dominated healthcare markets for regenerative medicine therapies. Regenerative Medicine. 2012; 7: 429–38.

127. RNAD Corporation. Avoiding the tragedy of the commons in health care policy options for covering high-cost cures. 2016. https://www.rand.org/pubs/perspectives/PE190.html.

128. Peacock SJ, Regier DA, Raymakers AJN, et al. Evidence, values, and funding decisions in Canadian cancer systems. Healthcare Management Forum. 2019; 32: 293–8.

129. Spinner D, Faulkner E, Carroll M, Ringo M, Joines J. Regenerative medicine and cell therapy in orthopedics-health policy, regulatory and clinical development, and market access. Techniques in Orthopaedics. 2019; 34: 224–43.

130. Yeung K, Suh K, Basu A, et al. Paying for cures: How can we afford it? Managed care pharmacy stakeholder perceptions of policy options to address affordability of prescription drugs. Journal of Managed Care & Specialty Pharmacy. 2017; 23: 1084–90.

131. Papadaki M. Adaptation through collaboration: Developing novel platforms to advance the delivery of advanced therapies to patients. Frontier in Medicine (Lausanne). 2017; 4: 56.

132. Schatz AA, Prejsnar KW, McCanney J, et al. Policy strategies for the "new normal" in healthcare to ensure access to high-quality cancer care. National Comprehensive Cancer Network. 2019; 17: 105–9.

133. Haute Autorité de santé (HAS). Medicinal products assessment-Principles of medicinal products assessment and appraisal for reimbursement purposes. 2018. https://www.has-sante.fr/jcms/c_2035649/en/assessment-of-medicinal-products.

134. Berger ML, Martin BC, Husereau D, et al. A questionnaire to assess the relevance and credibility of observational studies to inform health care decision making: An ISPOR-AMCP-NPC Good Practice Task Force report. Value in Health. 2014; 17: 143–56.
135. Bubela T, McCabe C. Value-engineered translation for regenerative medicine: meeting the needs of health systems. Stem Cells and Development. 2013; 22 Suppl 1: 89–93.
136. Yadav RK, Ali A, Kumar S, et al. CAR T cell therapy: Newer approaches to counter resistance and cost. Heliyon. 2020; 6: e03779.
137. Bandeiras C, Cabral JMS, Gabbay RA, et al. Bringing stem cell-based therapies for type 1 diabetes to the clinic: Early insights from bioprocess economics and cost-effectiveness analysis. Biotechnology Journal. 2019; 14: e1800563.
138. Firestone G. Immuno-oncology cell therapies: commercial considerations and strategies for the new decade. Cell and Gene Therapy Insights. 2020; 6: 798.
139. Bubela T, McCabe C, Archibald P, et al. Bringing regenerative medicines to the clinic: the future for regulation and reimbursement. Regenerative Medicine. 2015; 10: 897–911.
140. Toumi M, Rémuzat C, Thivolet M. Adaptive pathways may expand the gap between regulators and payers. Value in Health. 2015; 18: A574.
141. Lipska I, Hoekman J, McAuslane N, et al. Does conditional approval for new oncology drugs in Europe lead to differences in health technology assessment decisions? Clinical Pharmacology & Therapeutics. 2015; 98: 489–91.
142. European Medicines Agency. Parallel consultation with regulators and health technology assessment bodies. 2017. https://www.ema.europa.eu/en/human-regulatory/research-development/scientific-advice-protocol-assistance/parallel-consultation-regulators-health-technology-assessment-bodies.
143. Alliance for Regenerative Medicine. Getting ready for advanced therapy medicinal products (ATMPs) in Europe. 2019. https://alliancerm.org/sector-report/market-access-report/.
144. Leech AA, Neumann PJ, Cohen JT, et al. Balancing value with affordability: Cell immunotherapy for cancer treatment in the U.S. Oncologist. 2020; 25: e1117–e19.
145. CMS. gov. CMS issues final rule to empower states, manufacturers, and private payers to create new payment methods for innovative new therapies based on patient outcomes. 2020. https://www.cms.gov/newsroom/press-releases/cms-issues-final-rule-empower-states-manufacturers-and-private-payers-create-new-payment-methods.
146. Cutler D, Ciarametaro M, Long G, et al. Insurance switching and mismatch between the costs and benefits of new technologies. The American Journal of Managed Care. 2017; 23: 750–7.
147. Montazerhodjat V, Weinstock DM, Lo AW. Buying cures versus renting health: Financing health care with consumer loans. Science Translational Medicine. 2016; 8: 327ps6.
148. Leung K, Wu JT, Liu D, et al. First-wave COVID-19 transmissibility and severity in China outside Hubei after control measures, and second-wave scenario planning: a modelling impact assessment. The Lancet. 2020; 395: 1382–93.
149. World Health Organization (WHO). Coronavirus disease (COVID-19) pandemic. 2020. https://www.who.int/emergencies/diseases/novel-coronavirus-2019.
150. Calmels B, Mfarrej B, Chabannon C. From clinical proof-of-concept to commercialization of CAR T cells. Drug Discovery Today. 2018; 23: 758–62.
151. U.S. Food and Drug Administration. Updated information for human cell, tissue, or cellular or tissue-based product (HCT/P) establishments regarding the COVID-19

pandemic. 2020. https://www.fda.gov/vaccines-blood-biologics/safety-availability-biologics/updated-information-human-cell-tissue-or-cellular-or-tissue-based-prod uct-hctp-establishments.

152. Ortiz de Landazuri I, Egri N, Munoz-Sanchez G, et al. Manufacturing and management of CAR T-cell therapy in "COVID-19's time": Central versus point of care proposals. Frontiers in Immunology. 2020; 11: 573179.

153. Fierce Pharma. Novartis sidesteps Europe travel ban to provide CAR-T drug Kymriah to patients. 2020. https://www.fiercepharma.com/manufacturing/novartis-sidesteps-europe-travel-ban-to-provide-car-t-drug-kymriah-to-patients.

154. McKinsey&Company. COVID-19 and cell and gene therapy: How to keep innovation on track. 2020. https://www.mckinsey.com/industries/pharmaceuticals-and-medical-products/our-insights/covid-19-and-cell-and-gene-therapy-how-to-keep-innovation-on-track.

155. Elverum K, Whitman M. Delivering cellular and gene therapies to patients: Solutions for realizing the potential of the next generation of medicine. Gene Therapy. 2019; 27(12): 537–44.

156. VICTOR KOTSEV. Covid-19 forces gene therapy companies to shift strategy. 2020. https://www.labiotech.eu/gene-cell-therapy/covid-19-gene-therapy-shift/.

157. Aho J. How COVID-19 is changing the production of cell therapies. 2020. https://www.cellandgene.com/doc/how-covid-is-changing-the-production-of-cell-therapies-0001.

158. Aledo-Serrano A, Gil-Nagel A, Isla J, et al. Gene therapies and COVID-19 vaccines: A necessary discussion in relation with viral vector-based approaches. Orphanet Journal of Rare Diseases. 2021; 16: 316.

159. Psotka MA, Abraham WT, Fiuzat M, et al. Conduct of clinical trials in the era of COVID-19: JACC scientific expert panel. Journal of the American College of Cardiology. 2020; 76: 2368–78.

160. Kili S. Clinical trials in the era of Covid-19: Successes, failures & ongoing challenges. Cell and Gene Therapy Insights. 2020; 6: 775–82.

161. Amorosi D. COVID-19, manufacturing challenges limit cell and gene therapy progress, FDA official says. 2020. https://www.healio.com/news/hematology-oncology/20201028/covid19-manufacturing-challenges-limit-cell-and-gene-thera py-progress-fda-official-says.

162. GlobalData. COVID-19 Clinical trial disruption declines but trials continue to suffer from slow recruitment. 2020. https://www.globaldata.com/covid-19-clinical-trial-disruption-declines-trials-continue-suffer-slow-recruitment/.

163. U.S. Food and Drug Administration. Rare disease therapy development and access remain top FDA priorities during coVID-19. 2020. https://www.fda.gov/news-events/fda-voices/rare-disease-therapy-development-and-access-remain-top-fda-priorities-during-covid-19.

164. Frey N, Porter D. Cytokine release syndrome with chimeric antigen receptor T cell therapy. Biol Blood Marrow Transplant. 2019; 25: e123–e27.

165. Mooraj H, Kawalekar O, Gupta L, Shah S. Delivering scientific innovation requires operating model innovation. 2020. https://www2.deloitte.com/content/dam/insights/us/articles/6642-cell-and-gene-therapies/DI_Cell-and-gene-therapies.pdf

166. Mezher M. FDA officials, experts discuss impact of COVID-19 on cell and gene therapies. 2020. https://www.raps.org/news-and-articles/news-articles/2020/10/fda-officials-experts-discuss-impact-of-covid-19-o

167. COMMISSION USSAE. bluebird bio provides assessment of impact of COVID-19, update on business operations and clinical program development. 2020. https://www.businesswire.com/news/home/20200326005305/en/bluebird-bio-Provides-

Assessment-of-Impact-of-COVID-19-Update-on-Business-Operations-and-Clini cal-Program-Development

168. Bluebird Bio. bluebird bio reports third quarter 2020 financial results and highlights operational progress. 2020. https://www.businesswire.com/news/ home/20201104005691/en/bluebird-bio-Reports-Third-Quarter-2020-Financial-Results-and-Highlights-Operational-Progress.

169. Alliance for Regenerative Medicine. Advancing Innovation during COVID-19-ARM global regenerative medicine sector report: H1 2020. 2020. https://alliancerm. org/sector-report/h1-2020-report-pdf/.

170. Squibb BM. Bristol Myers Squibb provides regulatory update on lisocabtagene mara-leucel (liso-cel). 2020. https://news.bms.com/news/details/2020/Bristol-Myers-Squibb-Provides-Regulatory-Update-on-Lisocabtagene-Maraleucel-liso-cel/ default.aspx

171. Taylor NP. Sarepta cites 'overburdened' FDA as factor in DMD gene therapy delay. 2020. https://www.fiercebiotech.com/biotech/sarepta-cites-overburdened-fda-as-factor-dmd-gene-therapy-delay

172. Qiu T, Wang Y, Dabbous M, et al. Current state of developing advanced therapies for rare diseases in the European Union. Expert Opinion on Orphan Drugs. 2020; 8: 417–29.

173. Lorgelly PK, Adler A. Impact of a global pandemic on health technology assess-ment. Appl Health Econ Health Policy. 2020; 18: 339–43.

174. National Institute for Health and Care Excellence. Betibeglogene autotemcel for treating transfusion-dependent beta-thalassaemia [ID968]. 2020. https://www.nice. org.uk/guidance/indevelopment/gid-ta10334.

175. National Institute for Health and Care Excellence. Onasemnogene abeparvovec for treating spinal muscular atrophy type 1 [ID1473]. 2020. https://www.nice.org.uk/ guidance/indevelopment/gid-hst10026.

176. Leahy J, Hickey C, McConnell D, et al. Coronavirus disease 2019: Considerations for health technology assessment from the National Centre for Pharmacoeconomics Review Group. Value in Health. 2020; 23: 1423–6.

177. Kunz CU, Jörgens S, Bretz F, et al. Clinical trials impacted by the COVID-19 pan-demic: Adaptive designs to the rescue? Statistics in Biopharmaceutical Research. 2020; 12: 461–77.

178. Luxner L. COVID-19 delaying rare disease and gene therapy trials, Pharma Execs say. 2020. https://alsnewstoday.com/news/covid-19-delaying-rare-disease-gene-therapy-trials-pharma-execs-say/

179. Vandenberghe LH. COVID-19: Gene transfer to the rescue? Human Gene Therapy. 2020; 31: 605–7.

180. Ramezankhani R, Solhi R, Memarnejadian A, et al. Therapeutic modalities and novel approaches in regenerative medicine for COVID-19. International Journal of Antimicrobial Agents. 2020; 56: 106208.

181. Cancio M, Ciccocioppo R, Rocco PRM, et al. Emerging trends in COVID-19 treat-ment: Learning from inflammatory conditions associated with cellular therapies. Cytotherapy. 2020; 22: 474–81.

182. Arabpour E, Khoshdel S, Tabatabaie N, et al. Stem cells therapy for COVID-19: A systematic review and meta-analysis. Frontier Medicine (Lausanne). 2021; 8: 737590.

183. Cell and Gene Catapult. Cell and gene therapy GMP manufacturing in the UK: Capability and capacity analysis. 2020. https://ct.catapult.org.uk/sites/default/files/ publication/ManufacturingReport2020_PUBLISHED.pdf.

184. Moderna. Moderna and vertex establish new collaboration to treat cystic fibrosis using gene editing. 2020. https://investors.modernatx.com/news/news-details/2020/Moderna-and-Vertex-Establish-New-Collaboration-to-Treat-Cystic-Fibrosis-Using-Gene-Editing/default.aspx

185. U.S. Food and Drug Administration. FDA takes additional action in fight against COVID-19 by issuing emergency use authorization for second COVID-19 vaccine. 2020. https://www.fda.gov/news-events/press-announcements/fda-takes-additional-action-fight-against-covid-19-issuing-emergency-use-authorization-second-covid.

186. Cell&Gene. The lasting COVID-19 impact on cell and gene therapy. 2020. https://www.cellandgene.com/doc/the-lasting-covid-impact-on-cell-and-gene-therapy-0001.

187. U.S. Food and Drug Administration. Emergency use authorization. 2020. https://www.fda.gov/emergency-preparedness-and-response/mcm-legal-regulatory-and-policy-framework/emergency-use-authorization.

188. Papadaki M. Adaptation through collaboration: Developing novel platforms to advance the delivery of advanced therapies to patients. Frontiers in Medicine. 2017; 4: 56.

189. Bubela T, Bonter K, Lachance S, et al. More haste, less speed: Could public-private partnerships advance cellular immunotherapies? Frontiers in Medicine. 2017; 4: 134.

190. Martínez-Noya A, Narula R. What more can we learn from R&D alliances? A review and research agenda. BRQ Business Research Quarterly. 2018; 21: 195–212.

191. Alliance for Regenerative Medicine. Innovation in the time of COVID-19: ARM global regenerative medicine & advanced therapy sector report. 2020. https://www.pharmiweb.com/press-release/2020-08-10/new-report-regenerative-medicine-advanced-therapies-sector-thriving-despite-covid-19.

192. Goldstein DA, Sarfaty M. Cancer drug pricing and reimbursement: Lessons for the united states from around the world. Oncologist. 2016; 21: 907–9.

193. Ingusci S, Verlengia G, Soukupova M, et al. Gene therapy tools for brain diseases. Frontiers in Pharmacology. 2019; 10: 724.

194. Lee CE, Singleton KS, Wallin M, et al. Rare genetic diseases: Nature's experiments on human development. iScience. 2020; 23: 101123.

195. Yasuhara T, Kawauchi S, Kin K, et al. Cell therapy for central nervous system disorders: Current obstacles to progress. CNS Neuroscience & Therapeutics. 2020; 26: 595–602.

196. Colasante W, Diesel P, Gerlovin L. How are cell and gene therapies changing drug development models? In: cell&gene, ed., 2018. https://www.cellandgene.com/doc/how-are-cell-and-gene-therapies-changing-drug-development-models-0001.

197. ten Ham RMT, Hoekman J, Hövels AM, et al. Challenges in advanced therapy medicinal product development: A survey among companies in Europe. Molecular Therapy — Methods & Clinical Development. 2018; 11: 121–30.

198. Milne CP, Kaitin KI. Challenge and change at the forefront of regenerative medicine. Clinical Therapeutics. 2018; 40: 1056–9.

199. Kassir Z, Sarpatwari A, Kocak B, et al. Sponsorship and funding for gene therapy trials in the United States. JAMA. 2020; 323: 890–1.

200. Bock AJ, Johnson D. Regenerative medicine venturing at the university-industry boundary: Implications for institutions, entrepreneurs, and industry. In: Schmuck EG, Hematti P, Raval AN, eds., Cardiac Extracellular Matrix: Fundamental Science to Clinical Applications. Cham: Springer International Publishing, 2018.

201. California Institute for Regenerative Medicine (CIRM). CIRM-Funded Institutions. 2020. https://www.cirm.ca.gov/our-funding/funded-institutions.

202. Cell and Gene Therapy (Catapult). Advancing cell and gene therapies through powerful collaborations. 2022. https://ct.catapult.org.uk/.
203. Harrison R, Gracias A, Mitchell W. Translating regenerative medicine science into clinical practice: The local to global pivot. Cell and Gene Therapy Insights. 2018; 4.
204. Segers J-P. Towards a typology of business models in the biotechnology industry. 2017. https://papers.ssrn.com/sol3/papers.cfm?abstract_id=3065300
205. Kong X, Wan J-B, Hu H, et al. Evolving patterns in a collaboration network of global R&D on monoclonal antibodies. MAbs. 2017; 9: 1041–51.
206. Smith DM, Culme-Seymour EJ, Mason C. Evolving industry partnerships and investments in cell and gene therapies. Cell Stem Cell. 2018; 22: 623–6.
207. McKinsey&Company. Biopharma portfolio strategy in the era of cell and gene therapy. 2020. https://www.mckinsey.com/industries/life-sciences/our-insights/biopharma-portfolio-strategy-in-the-era-of-cell-and-gene-therapy
208. McKinsey&Company. Gene therapy coming of age: Opportunities and challenges to getting ahead. 2019. https://www.mckinsey.com/industries/life-sciences/our-insights/gene-therapy-coming-of-age-opportunities-and-challenges-to-getting-ahead.
209. Merck. Merck to collaborate with GenScript to accelerate cell and gene therapy industrialization in China. 2019. https://www.worldpharmanews.com/merckgroup/4757-merck-to-collaborate-with-genscript-to-accelerate-cell-and-gene-therapy-industrialization-in-china
210. Adams D, Russotti G, Lunger J, et al. Evolving autologous and allogeneic cell therapy manufacturing models in the commercial setting. Cell and Gene Therapy Insights. 2020; 6: 830–43.
211. Adner R. Match your innovation strategy to your innovation ecosystem. Harvard Business Review. 2006; 84: 98–107; 48.
212. Banda G, Tait J, Mittra J. Evolution of business models in regenerative medicine: Effects of a disruptive innovation on the innovation ecosystem. Clinical Therapeutics. 2018; 40: 1084–94.
213. Chang MDPD. Innovating CAR T cell therapy for today and tomorrow. Cell and Gene Therapy Insights. 2020; 6: 783–8.
214. Wu C-Y, Roybal KT, Puchner EM, et al. Remote control of therapeutic T cells through a small molecule-gated chimeric receptor. Science. 2015; 350: aab4077-aab77.
215. Grosser R, Cherkassky L, Chintala N, et al. Combination immunotherapy with CAR T cells and checkpoint blockade for the treatment of solid tumors. Cancer Cell. 2019; 36: 471–82.
216. Vaddepally RK, Kharel P, Pandey R, et al. Review of indications of FDA-approved immune checkpoint inhibitors per NCCN guidelines with the level of evidence. Cancers (Basel). 2020; 12.
217. Deloittes. 2020 global life sciences outlook-Creating new value, building blocks for the future. 2020. https://www2.deloitte.com/us/en/insights/industry/life-sciences/global-life-sciences-outlook-2020.html.
218. Fidler e. In a surprise, UniQure sells its hemophilia gene therapy for $450M. In: BIOPHARMADIVE, ed., 2020. Fidler e. https://www.biopharmadive.com/news/uniqure-csl-behring-hemophilia-gene-therapy/580507/#:~:text=In%20a%20surprise%2C%20UniQure%20sells%20its%20hemophilia%20gene,to%20CSL%20Behring%20for%20%24450%20million%20in%20cash.
219. EMERGEN RESEACH. Cell and Gene Therapy Market Size Worth USD 6.57 Million By 2027. 2022. https://www.emergenresearch.com/press-release/global-cell-and-gene-therapy-market#:~:text=Vancouver%2C%20B.C.%2C%20March%2031%2C%202020%20-%20The%20global,according%20to%20a%20new%20report%20by%20Emergen%20Research.

220. AuWerter T, Smith J, The L. Cell and Gene Therapy: Biopharma Portfolio Strategy. McKinsey, 2020. https://www.mckinsey.com/industries/life-sciences/our-insights/biopharma-portfolio-strategy-in-the-era-of-cell-and-gene-therapy.

221. Quinn C, Young C, Thomas J, et al. Estimating the clinical pipeline of cell and gene therapies and their potential economic impact on the us healthcare system. Value in Health. 2019; 22: 621–6.

222. Leclerc O, Suhendra M, The L. What are the biotech investment themes that will shape the industry? 2022. https://www.mckinsey.com/industries/life-sciences/our-insights/what-are-the-biotech-investment-themes-that-will-shape-the-industry.

223. Armstrong A. Novartis Eyes Next-gen AAV Gene Therapies in $1.75B Voyager Biobucks Pact. Fierce Biotech, 2022. https://www.fiercebiotech.com/biotech/novartis-eyes-next-gen-aav-therapies-175b-voyager-biobucks-pact.

224. BioMarin Up on Positive Gene Therapy Data in Hemophilia A. Fierce Biotech. 2021. https://www.fiercebiotech.com/biotech/biomarin-up-positive-gene-therapy-data-hemophilia-a.

225. Armstrong A. UniQure shares spike on Huntington's gene therapy data. 2022. https://www.fiercebiotech.com/biotech/unique-shares-spike-investors-see-bright-spot-12-month-huntingtons-gene-therapy-data

226. Armstrong A. LogicBio's Sun and Shares Rise Again after FDA Releases Clinical Hold on Pediatric Genome Editing Therapy. Fierce Biotech, 2022. https://www.fiercebiotech.com/biotech/logicbios-sun-rises-again-after-fda-releases-clinical-hold-pediatric-genome-editing-therapy.

227. Gene and Cell Therapy FAQ's. American Society of Gene and Cell Therapy, 2022. https://asgct.org/education/more-resources/gene-and-cell-therapy-faqs.

228. DePinto J, Richards R. This is not a sustainable model. Cell & Gene Therapy Insights. 2022; 8(2), 159–74.

229. McDonald H. Advancing patient access to cell & gene therapies: partnerships, pilots, & psyche. Cell & Gene Therapy Insights. 2022; 8(1): 43–51.

230. Khurana M, Singh S. Business model innovation in cell and gene therapy. 2022. https://www.oliverwyman.com/our-expertise/insights/2022/feb/business-model-innovation-in-cell-and-gene-therapy.html.

231. Mooraj H. Cell and Gene Therapies. Deloitte Insights, 2020. https://www2.deloitte.com/content/dam/insights/us/articles/6642-cell-and-gene-therapies/DI_Cell-and-gene-therapies.pdf.

232. Cao M. Manufacturing challenges & considerations for allogeneic cell therapy. 2022. https://hansonwade-intelligence.com/manufacturing-challenges-and-considerations-for-allogeneic-cell-therapy/

233. Shah S, Kumar R. Is Gene Therapy a Sustainable Business Model? Seattle, WA: Coherent Market Insights, 2018.

234. Teixeira A, Wu J, Luchi M, et al. From Lab to Marketplace, Succeeding with Gene Therapies. BCG Global, 2021. https://www.bcg.com/publications/2019/lab-marketplace-succeeding-gene-therapies.

235. May H, Petersen D. Understanding the business model of gene therapy companies—peeling the onion back to understand why time matters. LifeSciVC, 2021. https://lifescivc.com/2021/11/understanding-the-business-model-of-gene-therapy-companies-peeling-the-onion-back-to-understand-why-time-matters/.

236. Jaroslawski S, Caban A, Toumi M. Sipuleucel-T (Provenge(R)): Autopsy of an innovative change of paradigm in cancer treatment. Value in Health: The Journal of the International Society for Pharmacoeconomics and Outcomes Research. 2015; 18: A479.

237. Daniel A. Ollendorf JTCPJN. The right price: A value-based prescription for drug costs. Oxford University Press, 2021.

238. Suryaprakash S, Teixeira A, Choy M. The Changing Landscape for Cell and Gene Therapy. BCG Global, 2021. https://www.bcg.com/publications/2021/understanding-the-rapidly-changing-cell-and-gene-therapy-landscape.

239. Lambert J. Innovative payments for innovative therapies: adopting value-based models in the regenerative medicines & advanced therapy sector. BioInsights. 2019; 6(7): 941–5.

240. AuWerterT,SmithJ,SternbergJ,etal.Unlockingaccesstothegenetherapymarketinthe United States McKinsey, 2019. https://www.mckinsey.com/industries/life-sciences/our-insights/unlocking-market-access-for-gene-therapies-in-the-united-states

241. COMMUNIQUE DE PRESSE. Accord-cadre du 05/03/2021 entre le Comite economique des produits de sante et les entreprises du medicaments (Leems). Paris: CEPS, 2021. https://sante.gouv.fr/IMG/pdf/210305_-_cp_-_o.veran_o.dussopt_a. pannier-runacher_-_accord-cadre_ceps_-_leem.pdf.

242. Tunis S, Hanna E, Neumann PJ, et al. Variation in market access decisions for cell and gene therapies across the United States, Canada, and Europe. Health Policy (Amsterdam, Netherlands). 2021; 125: 1550–6.

243. Facey KA-O, Espin JA-O, Kent EA-OX, et al. Implementing outcomes-based managed entry agreements for rare disease treatments: Nusinersen and Tisagenlecleucel. https://www.research.ed.ac.uk/en/publications/implementing-outcomes-based-managed-entry-agreements-for-rare-dis.

244. Neyt M, Gerkens S, San Miguel L, et al. An evaluation of managed entry agreements in Belgium: A system with threats and (high) potential if properly applied. Health Policy. 2020; 124: 959–64.

245. Toumi M, Jarosławski S. Managed Entry Agreements and Funding for Expensive Therapies (1st ed.). 1 ed. Boca Raton: CRC Press, 2022.

246. OECD Health Working Paper No. 115: Performance-based managed entry agreements for new medicines in OECD countries and EU member states.

247. Harrison RP, Ruck S, Rafiq QA, et al. Decentralised manufacturing of cell and gene therapy products: Learning from other healthcare sectors. Biotechnology Advances. 2018; 36: 345–57.

248. Levitan B, Getz K, Eisenstein EL, et al. Assessing the financial value of patient engagement: A quantitative approach from CTTI's patient groups and clinical trials project. Therapeutic Innovation & Regulatory Science. 2018; 52: 220–9.

249. Benjamin K, Vernon MK, Patrick DL, et al. Patient-reported outcome and observer-reported outcome assessment in rare disease clinical trials: An ISPOR COA emerging good practices task force report. Value in Health. 2017; 20: 838–55.

250. ISPOR. ISPOR report: 21st annual European Congress. November 2018.

251. Boycott KM, Lau LP, Cutillo CM, et al. International collaborative actions and transparency to understand, diagnose, and develop therapies for rare diseases. EMBO Molecular Medicine. 2019; 11.

252. Courbier S, Dimond R, Bros-Facer V. Share and protect our health data: an evidence based approach to rare disease patients' perspectives on data sharing and data protection — quantitative survey and recommendations. Orphanet Journal of Rare Diseases. 2019; 14: 175.

Index